An introduction to

Experimental Physics

Colin Cooke

First published in 1996 by UCL Press

UCL Press Limited
University College London
Gower Street
London WC1E 6BT

and
1900 Frost Road, Suite 101
Bristol
Pennsylvania 19007-1598

The name of University College London (UCL) is a registered
trade mark used by UCL Press with the consent of the owner.

British Library Cataloguing in Publication Data
A catalogue record for this book is available from the British Library.

Library of Congress Cataloging-in-Publication Data are available.

ISBNs
1-85728-578-6 HB
1-85728-579-4 PB

Typeset in Times.
Printed and bound by
Biddles Ltd, Guildford and King's Lynn, England.

Contents

List of examples

Preface

The book is written primarily for first year undergraduates, though a well-motivated sixth-former will benefit from some sections. Worked examples are used throughout, and every encouragement is given to readers to perform experiments for themselves. The best way of learning experimental physics is to do it; using this book as a substitute for carrying out experiments will damage your education.

CHAPTER 1

What experimental science is about

```
Repeat
  Keep notes
  Design experiment
  Measure experimental variable(s)
  Analyze data
  Model experimental situation
  Compare model/data
Until comparison satisfactory
Write scientific paper
```

The topics discussed in this book are shown above related to each other in the style of a computer program. The essence of science is that concepts developed in the brain, such as hypotheses, theories or models, must be compared with the results of controlled experiments or observations of natural phenomena. Only when the comparison is satisfactory do scientists feel content, and then perhaps only temporarily, before new ideas or measurements disturb the status quo.

The word "science" derives from the Latin *scire*, to know. It may once have encompassed all knowledge, but is now a subset, albeit a vitally important one. To put the subject of this book in context we can divide learning into four categories:

- experimental science;
- observational science;
- quasi-science;
- non-science.

The distinctive element of experimental science is that we have a measure of control over the conditions in which observations are made. Physics is a prime example of a science where we are able to regulate such experimental variables as temperature, pressure, and so on. Millikan's determination of the charge on the electron, described in Chapter 3, is a fine example. This ability to control conditions is one reason why the subject has developed so rapidly in recent centuries.

Astronomy is an example of a science where measurements can be made, but we do not have much control over the source of the observations. Radiation from the Sun has been observed in great detail and at many wavelengths, so that we have gained a lot of information about its structure and nuclear chemistry, but we have to take it as it is. Theories should still be quantitative and tested against observations in order to be classified as science.

Quasi-science can be represented by subjects like psychiatry and sociology. Controlled experiments are virtually unknown, and though observations are clearly possible they cannot be tested against quantitative theories. Models can be made of how brains or societies work, but these do not have the required objectivity to be classed as science.

Art, music and literature have no pretence to be scientific, and are none the worse for that. Science and technology can help in the design of equipment for use in these activities but the content is artistic rather than scientific.

Before looking at the details of this approach to experimental physics it is worth asking what its strengths and limitations are. The former may need discussing in order to persuade you to continue reading, or even to rush out and buy your own copy of this book. You will need to invest a lot of time to study experimental physics properly, so you ought to be convinced of its importance. On the other hand it does no service to science to forget that it is a system created by human beings, so its application may share some of our shortcomings.

Strengths of the method

Over the last few centuries, experimental science, with its allied subjects of engineering and technology, has developed more rapidly than any other subject of concern to people. The greater understanding of the way the universe operates and the range of applications in everyday life have been immense.

In any activity involving human beings, disputes will arise. Scientists solve theirs by submitting them to the splendid method of test by experiment. You might care to compare this with methods used in activities such as religion, politics and law, to mention only three.

Since the method is applied by imperfect human beings it is almost inevitable that cheating will happen occasionally. The checks and balances implied in the experimental method usually bring these to light sooner rather than later. The same applies where genuine mistakes are made.

Although the personality of the experimenter is often important in the success of a piece of work, it does not intrude as much as in some other activities; objectivity is essential in science. Sometimes this requires the use of a special variation of scientific practice called the "double blind" method. This can be illustrated by considering what happens when the effect of a new drug is tested on patients. It is well known that some patients exhibit the placebo effect, i.e.

appear to improve when treated by something innocuous, merely because they expect to feel better. Conversely the doctor trying out the new drug may have a stake in its effectiveness and needs to be guarded against the temptation to interpret the data favourably. The double blind method uses people who form an information barrier between doctor and patient so that neither knows whether a particular individual has received the drug or placebo. Only when the response of each patient has been evaluated is the doctor told who received which treatment.

Many of the skills acquired in the study of experimental physics are applicable outside the laboratory. This is no "ivory tower" system with no relevance to the "real world".

Problems in applying the method

It takes considerable time to become proficient in the method, not least because the only satisfactory way of learning is by doing, particularly in an experimental subject like science.

The method is not readily understood by non-scientists. This may lead to misunderstanding, particularly when a scientist is unable to interpret the work for a general audience. There is an added problem when scientists fail to agree on a subject of concern to the public, such as the degree of global warming expected in the future.

The method is not suitable for all pursuits. Whereas it may help in the *analysis* of a painting or piece of music, it will add nothing to our enjoyment of the same.

There is no mention of ethics; the method is just as applicable to designing nuclear weapons as to discovering penicillin. It is up to the individual scientist, or those who place constraints on the work, to add a moral element to the system.

Experimental skills

One way of looking at experimental physics is to consider four general skills needed for its execution:

- experimental design (Ch. 3);
- collection of data (Ch. 4);
- analysis of data (Ch. 5);
- reporting on the experiment (Ch. 2).

Failure in any one of these will degrade any fine work you have done in the others. As the skills must be used in combination they are best learned together, though at any given time you may be practising only some of them. This book will assist you in this study, if you let it, and will include many worked examples. But before that let us take a general look at each of the skills in turn.

Experimental design

This is the full set of procedures that are needed to take a topic from initial vague question right through to a report of an experiment in a scientific journal. It may be a very simple idea, such as the one I posed to a young friend: "How many times can you fold a sheet of paper?" I encouraged him to think about the problem, theorize (i.e. guess!) what the number might be, and then try the experiment for sheets of different size. Conversely it may be one of the biggest questions of all – the origin and age of the universe. As if that problem is not colossal enough, we are not able to perform controlled experiments on the system, though one can go a long way by observing it carefully.

The full series of steps is discussed in Chapter 3. One of the most difficult for a beginner is the first, deciding what problem to investigate. Through lack of time and experience students rarely have much say in this, needing to progress through a series of experiments designed to teach them basic techniques. As experience grows, your involvement in choosing subjects to investigate will increase. The last step, writing a report, is also important since what is the point of doing good work if nobody hears about it?

Collection of data

Sometimes measurements show no variation when repeated, since they are either crudely made or involve simple counting. This is rarely the case in scientific experiments, and the variability of repeated measurements is an important characteristic of the data. For science to progress we must agree how to deal with this variability. Even an arbitrary method would be tolerable provided we all agreed to use it, since at least we would all be talking the same language. In fact we can do a little better than that, and I think you will agree that the method chosen – *the principle of least squares* – is also reasonable (see Ch. 5). You used it many years ago when first calculating an arithmetic mean, though you did not realize it at the time.

Crude measurements

A few occasions arise when only a crude measurement is needed, and barring mistakes there is no need to repeat it. If one were to do so then the same result would be obtained each time.

If, for example, I had a space measuring 1.5 m in which to place a desk 1.4 m long, then one measurement of each would suffice. Further, there is no need to measure either to an accuracy of much better than 0.1 m. The measurements would have been performed to an accuracy which is *appropriate for the purpose*. Any mistake would become apparent when the furniture was being arranged.

Simple counting

Another category of measurement in which variability is not to be expected is when we count a fixed number of objects. Again mistakes may be made, but competent people will soon identify and correct them. The importance of the measurement, and thus the care with which it is made, will still vary.

The counting may be performed to check that all the miners return from the coal face at the end of the shift who went to it at the beginning. A simple tally system is effective. Similarly, orienteering clubs must be sure that all competitors have returned from the woods, and are not lying injured somewhere. This is why it is advisable to carry a whistle to summon help should you sustain a serious injury and why you must report to the finish even if abandoning the event.

In the two previous cases the numbers in and out must tally *exactly*, but this will not be necessary at an election if one candidate has a large majority. It would be a poor loser who demanded a re-count after losing by 10 000 votes. Nobody cares whether the majority is 10 000 or 10 010. If the majority is small, however, then a re-count is certainly in order, as ten votes miscounted may make all the difference between winning and losing. Again, the quality of the measurement must be suited to its purpose.

How many measurements should we take?

Anyone can make a mistake in taking or recording a measurement, so it is wise to take another one if only as a check on the first.

Secondly, if a number of measurements of an experimental variable are made, under what are assumed to be the same conditions, we will not get the same answer each time. Although this variability will be smaller in a well controlled experiment than in one where the conditions fluctuate greatly, it is an important property of the measurement.

It is surely obvious that the more measurements you make, the more information you will have. But do not fall into the trap of collecting data with no thought for its quality; there is no point in having a small and well defined variability if it is swamped by a systematic error which means that all your readings are useless. There is no room in science for hard-working idiots.

Measurements with variability

We repeat a measurement a number of times, make our best endeavours to get the same value each time - and *fail*. The reason is that we have been unable to control the conditions of measurement perfectly each time. There could have been fluctuations in temperature, pressure, humidity, concentration, and so on.

Of course we must not fall into the trap of thinking that because conditions cannot be controlled perfectly then we need not make our best efforts to do so. Carelessness leads to greater variability in measurements, which will become apparent when comparing your results with those of someone who has taken more care.

Analysis of data

The appropriate statistics will be dealt with in Chapter 5 but an example would not come amiss now. Below are ten measurements of a distance, in metres:

2.00 2.03 1.98 2.01 1.95 2.02 2.01 1.97 2.01 1.99

How should we convey this information to another scientist?

The simplest answer is to quote all ten values and leave the user to make of them what he or she will. This might be practicable in this case, but what if there had been 1000 measurements? No one is going to thank you for such a lot of undigested information. What we need is some simple, and agreed, method of summarizing such data. Any summary will have the inevitable consequence of reducing the information transferred, but that defect will be repaid by the data being more usable. The complete set of measurements must be kept in your lab notebook in case anyone needs to see them all.

The dire need of our ancestors to distinguish the tiger from the grass has evolved in our brains a splendid ability to interpret patterns. Thus, we should think of displaying data in diagrams. Two types particularly useful in science are shown in Figures 1.1 and 1.2.

The first, a *bar chart*, is simply a graph with the measured values on the x axis and the number of occurrences on the y axis. It displays all the values, but loses information about the time or order in which they were collected. If there are a lot of measurements the shape of the graph may give useful information on the properties of the data collected. Should the data have a more continuous

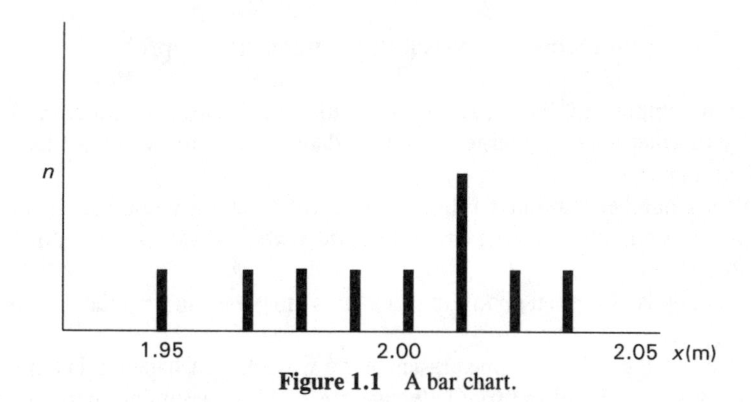

Figure 1.1 A bar chart.

nature than that shown it will be necessary to group values covering a small range, called a *bin*, and plotting this and adjacent ranges on the x axis. This is then referred to as a *histogram*. One has then lost the precise value of a given reading, knowing only that it is within the range of the bin. Again, this loss of information may be balanced by knowledge of the shape of the histogram.

The second type of diagram maintains the order of measurement, and is thus called a *time series* (Fig. 1.2). It is a graph with the values obtained on the y axis and the time of occurrence on the x axis. Since the times at which the measurements were taken is not known in the example given, we plot the order of their taking. Such a graph would be useful if you suspected that something unusual may have happened during the course of the measurements, in which case a discontinuous change, or a clear drift, would be apparent. No such trends appear in Figure 1.2.

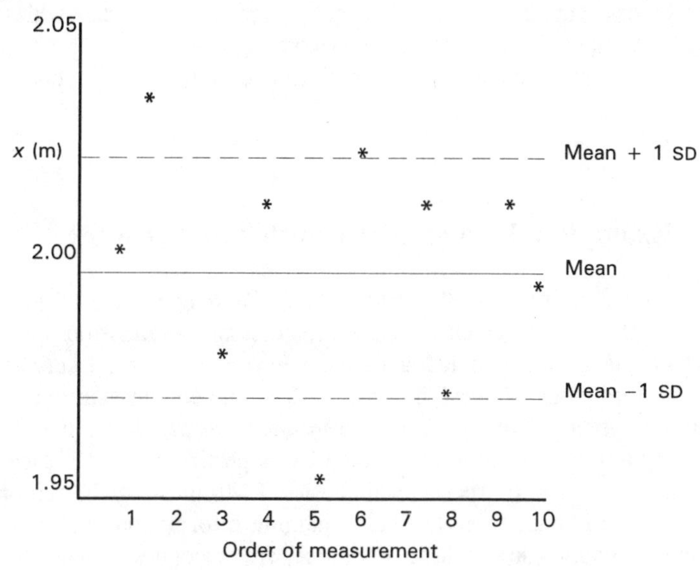

Figure 1.2 A time series (SD, standard deviation).

Measurements with variability – numerical approach

We need to supplement the visual approach discussed above with a numerical summary of what may be a large amount of data. There are two major uses of such a summary:

(a) to get a number that fairly reflects the "best" value to summarize the data, together with other numbers to indicate the spread in values about this best value;
(b) to distinguish between good and bad data, so that we can treat them accordingly.

For (a) we use the *arithmetic mean*, which we will see in Chapter 5 is the best estimate we can make of the *true value* we are seeking, assuming there are no *systematic errors* of significant magnitude. The latter are discussed later. The spread in our values is measured by the *standard deviation*, while the departure of the mean from the true value is measured by the *standard error of the mean*. Both are defined, discussed, and illustrated in Chapter 5.

We could invent many measures to satisfy (b) but there is no need to make another definition when the standard deviation will suffice. All other things being equal, a set of data with smaller standard deviation (small spread) will be of better quality. So if we combine two sets of data to get an overall mean, for example, we will give greater weight to the one having smaller standard deviation.

That glib phrase "all other things being equal" contains a lot of experimental physics, as it introduces the vitally important subject of systematic errors. Thus the ideal data set has both a small standard deviation and a small systematic error. The latter cannot be quantified except by comparing your data with someone else's, and if they disagree either one or both could be in error.

Some of the points discussed above are illustrated in the examples considered below.

Example 1.1 – a great scientific mind at work

On Tuesday, 28 January 1968, the space shuttle *Challenger* exploded just after take-off with the loss of seven lives. A Presidential Commission appointed to investigate the accident included Richard Feynman, Nobel Laureate and physicist extraordinaire. He describes his work on the commission in part two of his autobiography, *What do you care what other people think?* It is a book you are likely to enjoy reading, and some of it is pertinent to the present discussion of the principles of experimental physics. I will illustrate the approaches taken by Feynman under three headings: gaining information, analyzing the system, and communications. These, of course, are analogous to headings used above.

Gaining information

Feynman was not content simply to read reports as they were provided, but preferred to go out and talk to the people who knew most about the technical problems. Any filtering process which could easily occur in a large organization was thus avoided.

One of the crucial pieces of information which he was able to contribute to the commission was the effect of low temperatures on the loss of resilience in rubber O rings used as seals between the engine gases and the atmosphere. He eventually concluded that there was probably information on the subject in some NASA report, but for some reason it was not generally available. Another commission member, General Kutyna, claimed to have thought of the idea when working on the carburettor of his car, though he might have known about the possible existence of the NASA report. Feynman took up the problem and showed, by placing an O ring in iced water, that it did indeed become less flexible, certainly over the timescale of fractions of seconds over which the rocket expanded. Temperature was an essential parameter, as on previous flights the lowest temperatures encountered by the O rings was 11.5°C whereas *Challenger* had been launched at an ambient temperature of -1.5°C.

The problem with low temperatures appeared to be even worse when it became known that some measurements had been made with a thermometer not used under the conditions specified by its manufacturer. This is always likely to lead to systematic errors in the measurement, so it was necessary to recalibrate it for the conditions of actual use. When this was done the inconsistencies disappeared, though the low launch temperature of -1.5°C was still valid.

One useful source of information came from some of the 100 cameras used to look at the system during launch. These showed a small leak at the level of the O ring seals which may have been the precursor of the explosive leak that occurred a short time afterwards.

Parts of systems were reused for later flights, so that it was necessary to check the cylindrical sections for roundness. Feynman was amazed to find that these checks were made by measuring three diameters only. Such naivety is surprising, but we are all capable of over-simplification at times.

Interpretation of information

An earlier explosion on an unmanned Titan rocket had provided warnings that the cause of such accidents could be much more complicated than was thought at first sight. There had been less instrumentation in that case so there were fewer clues to go on, which meant that the investigating team changed its mind three times about the cause of the explosion.

Among the various discrepancies which appeared between the views of engineers and management on the project were the estimates made of the probability of an engine failure. Managers used a figure of 1 part in 100 000 while engineers thought a few parts in 100 was a more reasonable estimate. Feynman points out that the managers' estimate meant that there could be one flight per day for about 300 years before a failure occurred. This was way out of line with the problems that had arisen on earlier flights.

Communication of information

Some jargon is inevitable in any profession, but there ought to be some sensible limit, so that an intelligent outsider has a hope of following discussions. Feynman was astounded to find that NASA had a dictionary of acronyms that came as a big fat book. To give you a flavour of the jargon, try interpreting this: "The SSME's burn LH and LOX which are stored in the ET". With time to spare, and with only a single sentence to interpret, you might be able to find what the writer is saying, but imagine being bombarded with pages of such indigestible jargon. I wonder if the spell-checking software at NASA contains these acronyms - mine is not impressed by them! Meaning can be difficult enough without the use of acronyms, as evidenced by this sentence in a NASA report on turbine blades: "4000 cycle vibration is within our data base".

Poor communication between management and engineers was apparent. The former claimed not to know what the latter were thinking, while the engineers considered that their suggestions were ignored often enough for them to stop making any more. Not that such problems are only found in NASA; good communications are difficult to maintain in any large organization.

Later, I shall try to persuade you of the necessity to keep constant records of your experimental work, so it is worth pointing out that Feynman was writing reports as he went along. This has the double advantage of reducing the storage requirements of the brain and also providing a good communication link with others.

What may well surprise you is that there were 23 versions of the final report before it was regarded as satisfactory. Perhaps the ready availability of editing facilities in word-processing software encourages some of these modifications, and Feynman notes that in the time of typewriters there might only have been three versions. Technology does not guarantee progress.

Example 1.2 – problems with publishing results

Science is a social activity in that you must communicate your results and methods to others so that, if inclined, they are able to check and extend the work. The normal procedure is to submit a report of an experiment to the editor of a journal of your choice, who then sends copies to two scientists considered competent to review the paper. They then report back to the editor on the content and clarity of the report, and give their opinion on its suitability for publication. Often they will suggest changes that should be made before publication proceeds. The author's identity is known to the reviewers, but not the reverse; editors probably think that reviewers who could not preserve their anonymity would either not undertake the work or be inhibited in their response. I would prefer a symmetrical arrangement where author and reviewer were either both known to each other, or both not known.

Compare this system with the announcement of the discovery of cold fusion by Pons and Fleischmann on 23 March 1989. It must have seemed strange to many scientists that such an important discovery should be reported first in the financial press rather than in a scientific journal. In the latter it would have been subject to review by scientific peers, who would surely have challenged some of the data presented, and suggested further tests that should be made. But the importance of a cheap energy source was such that many groups of scientists dropped their current experiments to try to duplicate the results. In the main they failed to do so, and most people concluded that the original results were faulty in some respect. In his book, *Too hot to handle*, Frank Close questions whether respect for science is diminished by the unsatisfactory nature of this story. I believe the opposite to be true, in that it is a great achievement for scientific methods that the truth should be so quickly reached. It may serve as a warning of the pressures that commercialism can have on the scientific process.

Physicists do not have to consider the feelings of their equipment when describing an experiment; it may have different characteristics when you return after a lapse of time, but not because you have written about it! However the psychoanalyst's "equipment" is a patient, who would be justifiably distressed if personal details were made available to anyone who could afford the price of the journal in which their condition was described. Certainly the psychoanalyst who made such disclosures would not expect to attract many patients in the future. A simple solution is to disguise the identity of the patient when reporting the case, but then no one can check the diagnosis, treatment or its efficacy. The common medical procedure of asking a colleague for a second opinion is still not equivalent to the openness expected in science. One of the early cases described by Freud illustrates the problem. Although referred to as *Anna O*, she was later identified as Bertha Pappenheim so that others could make judgements of the claimed cure. It did not stand up to scrutiny (see *Freud – the man and the cause*, by Ronald W. Clark, p. 100).

Example 1.3 – experimenting with people

One has no qualms about experimenting with inanimate objects, but living systems are more restricting in what we should do to them. Nevertheless some interesting results can be obtained. An American psychologist, Stanley Milgram, performed an experiment on human behaviour that was reported in his book *Obedience to authority*. A summary will give a flavour of his results.

One can talk at length about the legal and philosophical aspects of obedience from many standpoints, but a scientific approach is to observe specific instances, or better still perform an experiment to see how people react in practice. The experiment to be described was first performed at Yale University and later repeated with more than a thousand participants at several other universities. These repetitions are necessary to ensure that conditions applicable in the first experiment did not make its results special to one place or a few people. Science must be universal.

Two people are invited to a psychology laboratory, ostensibly to assist in a study of memory and learning. One of them, the learner or "victim", is an actor in cahoots with the experimenter. He is seated in a chair, arms strapped to prevent excessive movement, and an electrode attached to a wrist. He is told to learn a list of word pairs, and mistakes will be punished by receiving electric shocks of increasing intensity. In fact he receives no shocks at all, but pretends to do so.

The second assistant, the "teacher", is the focus of the experiment. After watching the learner being strapped into place he is taken to the next room and sat in front of an impressive shock generator. The main feature is a set of switches labelled from 15 volts up to 450 volts, with verbal descriptions ranging from *slight shock* to *danger – severe shock*. The teacher is told to test the person in the other room. A correct response is to be treated by continuing to the next item; a mistake by giving an electric shock, starting at the lowest level and increasing on each occasion.

Conflict between teacher and experimenter usually begins when the learner feigns discomfort at low voltages. At 150 volts he demands to be released from the experiment, and protests become more vehement, until at 285 volts it becomes an agonized scream. Should the teacher respond to this apparent pain by stopping the experiment, or continue to bow to the wishes of the experimenter who enjoins him to continue after every hesitation? Can he make a break with an authority he has come to help?

You are sure to ask why anyone in his right mind would administer even the first shocks? But no one ever takes the simple expedient of walking out of the laboratory. Starting the procedure is perhaps understandable because the learner has appeared co-operative, if a trifle apprehensive. What is surprising is how far people will go in complying with the experimenter's instructions; a substantial proportion continue to the last shock on the generator. The extreme willingness of adults to go to almost any lengths on the command of an authority is the chief finding of the experiment.

CHAPTER 2
Writing about experiments

Use not vain repetitions

Matthew, Chapter 6, verse 7

In spring, when woods are turning green,
I'll try and tell you what I mean

Lewis Carroll, *Through the Looking Glass*

The shortest correspondence on record was that between Victor Hugo (1802–85) and his publisher Hurst and Blackett in 1862. The author was on holiday and anxious to know how his new novel, Les Miserables, was selling. He wrote, "?". The reply was "!"

Guinness Book of Records

As with other aspects of experimental physics, the best way to learn about communicating with others is by practising it. This takes time, thought, and a skill in the use of language that only comes "naturally" to a few. The rest of us have to work hard at it, though encouragement may be found in Samuel Johnson's observation that "what is written without effort is in general read without pleasure". In order to make the necessary commitment you need to be persuaded of the importance of the subject, and if in doubt consider the following points.

First, what is the sense of spending a lot of time, effort and imagination performing an experiment if you are not to gain credit for the work? You do this by communicating your work to other scientists, not by keeping it to yourself.

Secondly, one of the necessities of science is that your work be checked by other scientists. Secrecy may be tolerated in other aspects of life, but there is no place for it in science. Of course it is up to the individual to decide *when* communication will take place, and until that time there may be reluctance to present preliminary work or ideas, but communicate you must at some stage.

Thirdly, a good way of appreciating scientific writing is to read some of the excellent books that have been written on various aspects of the subject. A list of such books that I have found interesting is given in Appendix 3.

If you are still in need of inspiration, recall that the young Faraday was appointed to his job as assistant to Sir Humphrey Davy after attending lectures at the Royal Institution and sending a copy of his notes to Sir Humphrey.

In case you are not sure what a written report of an experiment should look like I have provided two versions below. Both discuss the same experiment, but the second is a complete report, in so far as any report is ever complete, and the first is a version with items omitted deliberately. Read the incomplete version first and consider what is missing from it. After that look at the full version to see what I think it should contain. Language is such a flexible tool that there cannot be a single definitive report of any experiment, and I am not foolish enough to consider mine as such. The intention is to give you some ideas on which to base your own reports.

Example 2.1 – reports of an experiment

These come in two forms: the one written for yourself in your lab notebook, and the one written for someone else to read. The second is based on the first, of course, but should take into account that the reader may not be as familiar with the subject as you are, now. Modesty will forbid you including all the details contained in your notebook, even if space does not. The selection you then make must be a fair reflection of the experiment without directing the reader down every blind alley you have stumbled into. You must try to consider the needs of the readers or they may miss the gems you are presenting.

The photoelectric effect (report with deliberate omissions)

Introduction

In the normal operation of a photoelectric diode the anode is maintained at a positive potential with respect to the photocathode so that the electrons will be collected. In this experiment, however, the anode potential is made steadily more negative. This discourages the electrons from flowing towards the anode, until eventually they stop doing so because the electric field opposing them overcomes their original kinetic energy. The voltage which just achieves this object, V_{max}*, is a measure of the maximum kinetic energy of the electrons leaving the cathode.*

Method

The arrangement of the apparatus is shown in Figure 2.1 and circuit diagram in Figure 2.2. The source of light was a quartz iodine lamp, filtered so that only a small range of wavelengths reached the cathode. Preliminary measurements were made to discover which range of the electrometer was suitable for use with each filter.

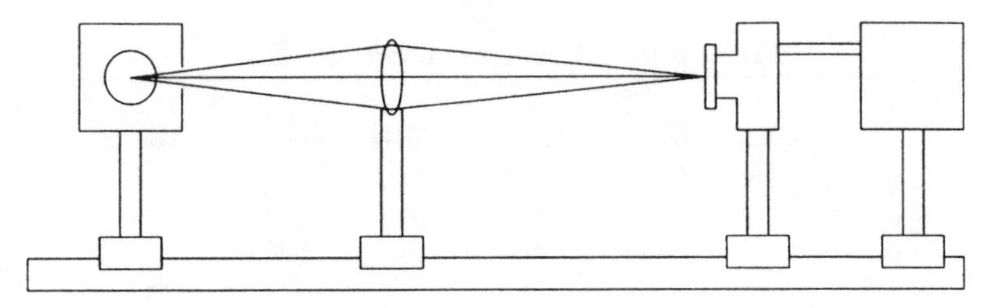

Figure 2.1 The apparatus on an optical bench.

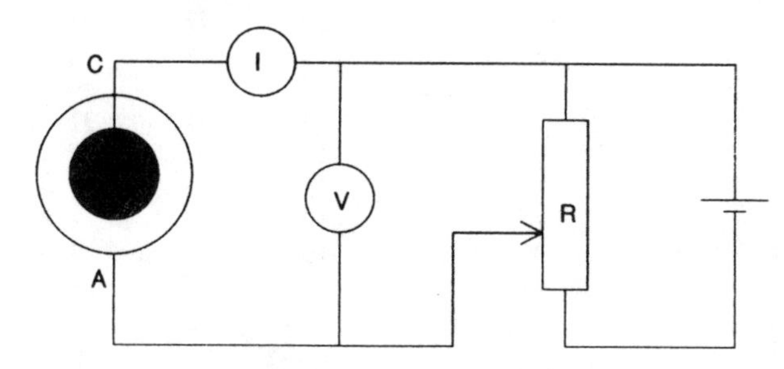

Figure 2.2 The electrical circuit.

Starting with a potential difference of 0 V, and incrementing in 0.1 steps, the current was read for each filter in turn. To reduce the possibility of drift in the readings, the measurements with all six filters were taken in the shortest time consistent with accuracy.

A single sheet of graph paper was used to plot all six current/voltage curves. This enabled easy comparison to be made between the results for different filters, and enlarged graphs could be drawn later if greater sensitivity was needed.

Results

Table 2.1 contains the data for all six filters, and the graphs are shown in Figure 2.3. Voltages were measured to 0.01 V, and currents to 1 or 2 units.

Analysis

A graph of V_{max} against $1/\lambda$ will yield a straight line.

In spite of all attempts to prevent light reaching the anode, some electrons were emitted from it, yielding a negative current that was evident when the voltage was large enough. This made it more difficult to decide the value of V to choose as V_{max}, i.e. when the cathode current, I, was just zero. The data are shown in Table 2.2, to which is added the analysis of the straight line, shown graphically in Figure 2.4.

Experimental value of slope $= 1.28 \times 10^{-6}$

The agreement between theory and experiment is reasonable.

$A = 1.78$

Table 2.1 Values of the cathode current *I*.

	λ					
V	651.6	600.9	550.1	499.4	449.8	401.2
0.0	29	339	1145	1240	812	450
0.1	7	170	790	970	676	395
0.2	–4	71	464	706	540	341
0.3	–7	13	215	440	417	288
0.4	–8	–8	79	260	300	239
0.5	–9	–15	20	130	206	190
0.6	–9	–17	–5	56	127	146
0.7		–18	–15	20	72	106
0.8		–18	–20	3	37	75
0.9			–22	–8	18	50
1.0			–23	–14	6	31
1.1			–23	–16	–2	17
1.2				–18	–6	9
1.3				–19	–9	3
1.4				–19	–11	–3
1.5					–12	–5
1.6					–12	–7
1.7						–9
1.8						–10
1.9						–10
2.0						

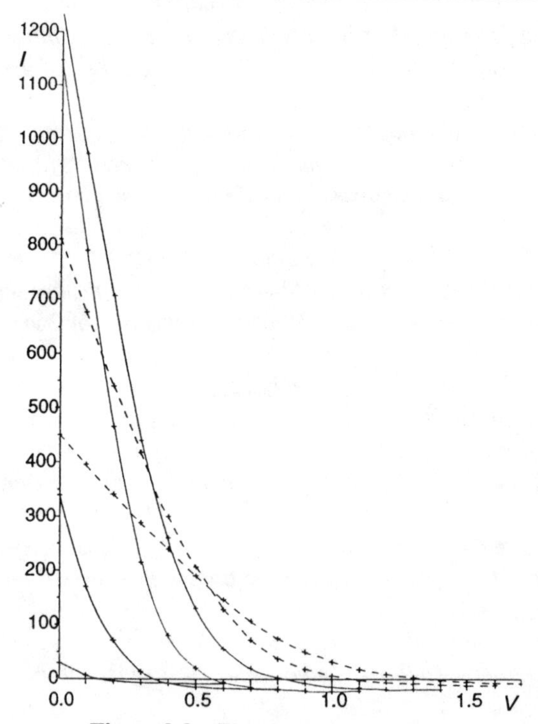

Figure 2.3 The photoelectric effect.

Table 2.2 The photoelectric effect.

λ	$1/\lambda(10^{-3})$	V_{max}	dV_{max}
651.6	1.535	0.14	0.01
600.9	1.664	0.36	0.01
550.1	1.818	0.57	0.01
499.4	2.002	0.82	0.01
449.8	2.223	1.04	0.02
401.2	2.493	1.32	0.02

Slope = 1.28.
Intercept = 1.78.

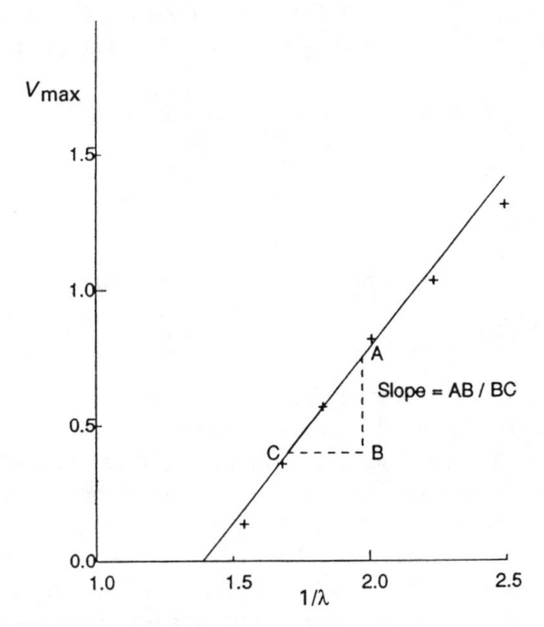

Figure 2.4 Plot of V_{max} versus $1/\lambda$.

Discussion

Perhaps the least satisfactory aspect of this experiment lies in the criterion used to choose the value of V_{max} from the measured data. An improvement may be achieved by concentrating only on those values of current close to zero, as there is little useful information at larger values.

Conclusions

(a) *Figure 2.4 is a straight line graph, thus verifying Einstein's photoelectric equation.*
(b) *The slope has a value 1.28;*
A is found to be 1.78.

Measurements on the photoelectric effect (complete report)

Comments on what the incomplete report should have contained will be made at the beginning of each section, using bold type. This will be followed by the complete report in the usual type.

There are two different types of report that you will have to write. The first is that in your lab notebook, which is primarily to remind you what happened during the experiment, and the second is a more formal account written mainly for other scientists to read. I will refer to these as the lab notebook and scientific reports. Since they are both describing the same experiment they will obviously have many things in common, but there are also important differences that we must consider. The contents are:

Lab notebook	Scientific report
	Abstract
Aims	Aims
	Introduction
Method	Method
Results	Results
Anaylsis (including errors)	Analysis (including errors)
	Discussion
Conclusion	Conclusion

Before discussing these headings in detail there are some general points to be made. The report in your lab notebook should be written *as the experiment proceeds* without any recourse to memory, however good you believe yours to be. Also, it is perfectly satisfactory to use numbered notes to describe each point. This will make it easier to retrieve the information needed to write your scientific report later. The latter must be written in clear English as it is to be read by others, which is not to say that your lab notebook can be scrappy. If it is, the day will surely come when you cannot read your own writing! Other people may also need to refer to it, if your scientific report has not given them all the information they need. The three extra items in the right-hand column above are put there specifically for the reader of your work, though you may decide to make notes on them in your lab notebook.

Abstract

This is a short summary of the whole experiment. You might wonder why this should be needed, since the report contains the same information in expanded form. The reason is that the scientific literature is so extensive that it is very difficult to keep in touch, even with the small part that is of interest to you. Thus a system is needed that will filter out those reports that are not of sufficient concern to justify reading in full.

The first filter comes with the title of the report. If this is written carefully it will tell the reader whether the subject is likely to be of interest (note the small but significant change in the title of the second report, which conveys its experimental rather than theoretical nature). If so, the next step is to read a summary of the report, called an abstract, which will convince the reader whether or not to spend a much longer time reading the whole report. As a guide you can think of the abstract as being 100–200 words in length.

In my example I have briefly described the measurements which have been made, the analysis of them, and the conclusions reached. There is not space to elaborate, but the abstract should be a fair summary of the full report.

Light from a quartz iodine lamp was filtered to give a narrow range of wavelengths, centred about a mean value λ, to focus on the cathode of a photodiode. An increasingly negative voltage was applied to the anode until, at a value V_{max}, electrons were just prevented from reaching it. A graph of V_{max} against $1/\lambda$ was a straight line, as expected, with a slope of $(1.280 \pm 0.018) \times 10^{-6}$ V m compared with the theoretical value (hc/e) of 1.240×10^{-6} V m. The work function of the cathode was found to be 1.781 ± 0.034 eV.

Aims

You are likely to have at least a rough idea what you are trying to achieve in performing an experiment, though students have been known to carry out an investigation by reading the script as they go along, rather than getting an overall picture before starting. Half an hour spent in planning can save hours later, as you will then know precisely what steps are needed and have a clear idea how to achieve your aims. Do not forget to discuss the experiment with someone who has already performed it. This is not cribbing but good scientific practice if carried out in the correct spirit.

Think of your report as an inverse sandwich, that is, one in which the significant parts are at the extremities: *aims* at the beginning and *conclusions* at the end. The first poses the questions and the second answers them.

The aims were omitted from the first account to see how you would react to an unfocused experiment. If your intentions are vague, how can you expect your reader to understand? Of course it may gradually become obvious as more of the report is read, but there is no substitute for stating clearly and concisely at the beginning what you intend to do. There will be times when you modify your plans during the experiment, which is fine provided you state clearly your revised aims.

(a) To verify the validity of Einstein's equation, which equates the energy carried by a photon with the sum of the energies required to extract the electron from the surface and the kinetic energy when released:

$$hf = A + \frac{1}{2}mv^2 \qquad (1)$$

where $$eV_{max} = \frac{1}{2}mv^2 \quad (2)$$

and $$c = f\lambda \quad (3)$$

(b) To obtain values for A and hc/e, and compare them with the expected ones. The symbols are explained in the text.

Introduction

What goes into this section depends strongly on who you expect to read the report, since its function is to give background information about the experiment. In many cases you would be well advised to assume the reader to be a student of similar experience to yourself who has not yet performed the experiment being described. The more unusual the experiment the longer would be your introduction. It might include a discussion of why you were doing the experiment, the physics underlying it, and perhaps what others have discovered about the subject. If you want to judge the quality of your introductions try swapping reports with another student and criticizing each others work; kindly but firmly!

It is impossible to write an introduction at a level to suit all readers. In the incomplete report, I made what was probably an unjustified assumption that the reader would be familiar with the photoelectric effect, i.e. the manner in which electrons are emitted from a surface when a beam of electromagnetic radiation strikes it. This time I have assumed that you know little about that, but have a general knowledge of the wave and quantum concepts of light. To those for whom this choice is not appropriate, my apologies.

Before Einstein applied his mind to the problem, it was thought that the speed with which electrons were emitted from a surface, illuminated with electromagnetic radiation, would be directly related to the intensity of the beam. This would be in accord with the previously successful wave theory of light, since a stronger beam consists of greater electric and magnetic fields. The bigger the fields to "kick" electrons out of the surface, surely the faster they would move.

But the "obvious" solution is not necessarily the correct one in physics: it is the particle, not the wave nature of the beam that must be considered when the interaction occurs. If a photon has sufficient energy (hf) it will overcome the electrical attraction (A) of the surface for the electron and cause it to be emitted. Any extra energy possessed by the photon above this basic minimum will give the electron kinetic energy ($\frac{1}{2}mv^2$) away from the surface. If another electrode is placed close to the photoemitting surface then the flow of electrons can be controlled by the potential difference between the two.

In the normal operation of a photoelectric diode the anode is maintained at a positive potential with respect to the photocathode so that the electrons will be collected. The magnitude of the current is then directly related to the intensity of the beam striking the photocathode, an idea used in the light meters beloved of photographers and cricket umpires. In this experiment, however, the anode potential is made steadily more negative. This discourages the electrons from flowing towards the anode, until eventually they stop doing so because the electric field opposing them overcomes their original kinetic energy. The voltage which just achieves this object, V_{max}, is a measure of the maximum kinetic energy of the electrons leaving the cathode.

Method

This is the section in which you describe how the experiment was performed: what measurements were made, the precautions taken to get reliable results, etc. If you are working from a script written by someone else remember that the perfect script has not been written any more than the perfect report. Be prepared to think for yourself, but be modest enough to ask for advice or a second opinion. Never forget that science is a social activity, and mutual assistance is helpful for both parties. In any experiment I would expect a student to have thought about systematic errors that may have occurred; at least a discussion of their origin, but preferably measures taken to avoid or minimize them. It might, for example, be something as simple as switching on electrical apparatus in advance of its use, so that it can reach thermal equilibrium. And if you have done it, say it, so that you get the credit!

Experimental physics can be an unforgiving activity, in that you can do 99% of things right but be let down by the 1% you forget or get wrong. A lot of the method was described in the first report, but I left out some of the important experimental detail so that you would be encouraged to consider the minutiae of performing and reporting on experiments. The labelling of the diagrams was poor also. Some of the ideas can be expressed in general form and are thus applicable to many experiments: attainment of thermal equilibrium, preliminary checks on apparatus, thoughts about systematic errors, etc. Do not worry about making slips, provided you learn from them afterwards, and try not to blame the apparatus for your mistakes!

The arrangement of the apparatus is shown in Figure 2.5 and the circuit diagram in Figure 2.6.

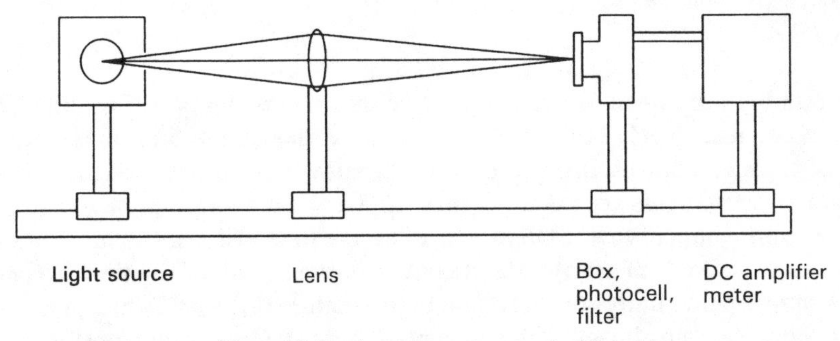

Figure 2.5 Apparatus on an optical bench.

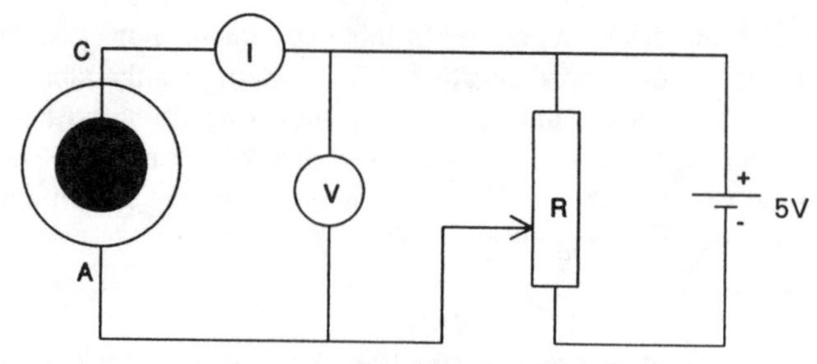

Figure 2.6 Circuit diagram.

The photodiode was shielded from stray light except at the window through which the beam entered. Preliminary tests showed that the photocurrent was negligible for the ambient light level in the laboratory. The source of light was a quartz iodine lamp, filtered so that only a small range of wavelengths reached the cathode at any time.

The apparatus was switched on straight away, to give time to reach thermal equilibrium while preliminary measurements were made. These were performed to discover which range of the electrometer to use with each filter; it was not thought desirable to change the range during a run with a particular filter as the meter calibration might change. Then the stability of a given reading was established over a timescale (about 5 min) similar to that used for a series of measurements with one filter.

Starting with a potential difference, V, of 0 V, and incrementing in 0.1 V steps, the current, I, was read for each filter in turn. To reduce the possibility of drift in the apparatus, the measurements with all six filters were taken in the shortest time consistent with accuracy. This took less than half an hour.

All six current/voltage graphs were plotted in a single diagram. This enabled easy comparison to be made between the results for different filters, and enlarged graphs could be drawn later if greater sensitivity was needed.

The apparatus was left switched on while the analysis of the data was made, in case further measurements became necessary. As a precaution should more readings be needed at a later date, each piece of apparatus was identified so that the same set could be used.

Results

The numbers obtained as the result of your measurements will usually be presented most clearly in the form of a table, with symbols and units at the top of each column. A little foresight will allow you to prepare the table before measurements are taken, so they can be written down immediately in a clear and compact form. Often you will have to transform the basic data in order to make it physically significant (e.g. the *square* of the time period for a simple pendulum is proportional to its length). It is worthwhile preparing enough columns in your table to contain both the basic and transformed

data, as there is nothing worse than having to keep turning pages to compare the two. To avoid doing this, or preparing another table, it is good practice to leave a couple of columns empty, in case you feel the need to carry out some other transformation not thought of at the time. Corrections should be made by crossing out the wrong value and writing in the correct one, not by overwriting, as this can become unclear. Neatness is desirable, clarity essential. Ambiguity may arise if you place any powers of ten alongside the units at the head of a column, since it is not always obvious whether the number has already been multiplied by 10, or is to be done so by the reader. Placed alongside, the number leaves no doubt of your intentions.

Every time you write the result of a measurement you should be thinking of three components: value, unit, and random error. Omissions were made from the first account to try to emphasize this fact. While it is a good idea, as in Figure 2.7, to show all six curves on the same sheet of graph paper, the absence of identification for the lines can lead to confusion. In your lab notebook use coloured pens to make them distinct.

Table 2.3 contains data for all six filters, with the corresponding plots shown in Figure 2.7.

Voltages were measured to 0.01 V, and currents to 1×10^{-12} A.

Table 2.3 Values of the cathode current I (10^{-12} A).

	λ(nm)					
V(V)	651.6	600.9	550.1	499.4	449.8	401.2
0.0	29	339	1145	1240	812	450
0.1	7	170	790	970	676	395
0.2	-4	71	464	706	540	341
0.3	-7	13	215	440	417	288
0.4	-8	-8	79	260	300	239
0.5	-9	-15	20	130	206	190
0.6	-9	-17	-5	56	127	146
0.7		-18	-15	20	72	106
0.8		-18	-20	3	37	75
0.9			-22	-8	18	50
1.0			-23	-14	6	31
1.1			-23	-16	-2	17
1.2				-18	-6	9
1.3				-19	-9	3
1.4				-19	-11	-3
1.5					-12	-5
1.6					-12	-7
1.7						-9
1.8						-10
1.9						-10
2.0						

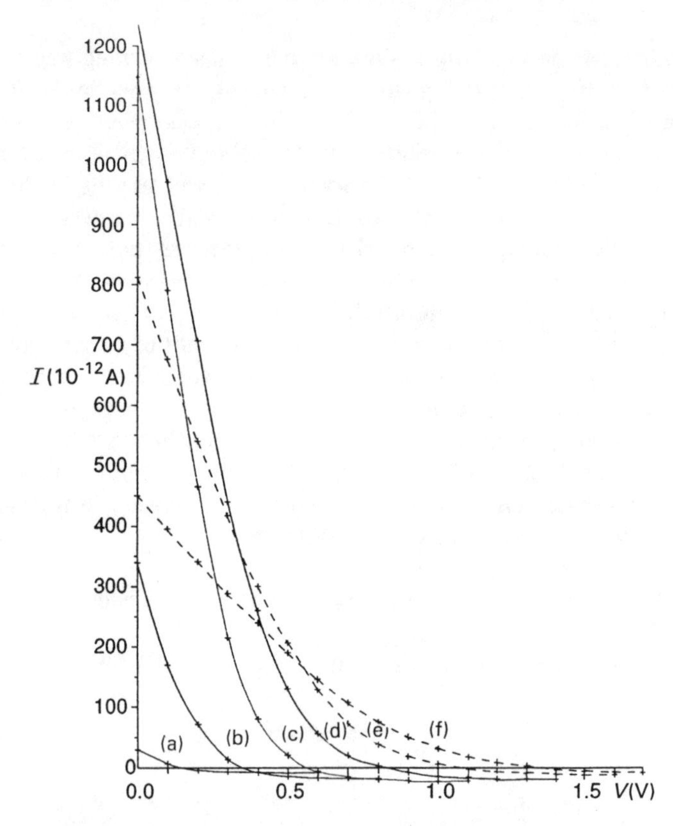

Figure 2.7 Voltage versus measured current at different wavelengths: (a) 651.6 nm, (b) 600.9 nm, (c) 550.1 nm, (d) 499.4 nm, (e) 449.8 nm and (f) 401.2 nm.

Analysis

It is a paradox that a subject like science, which deals in probabilities rather than certainties, has progressed much more rapidly that other activities which are less modest in their claims, though poorer in performance. Every report you write should include an estimate of uncertainty (called random error) in the final result. Also, you must develop the habit of considering possible systematic errors in experiments. These topics will be discussed in some detail later, but you should not consider a report complete without consideration of both.

Often data will submit to graphical analysis, and if a straight line is produced so much the better, as this is reasonably easy to interpret using statistics; you only need two constants to fit a straight line, three or more for a curve.

It is wise to make a clear distinction between measured data and the deductions made from them. Then if there is a problem of interpretation in your experiment it will be easier to target the weak spot.

The linear equation used to analyse the data should have been introduced as in the second account, not simply stated as in the first. By omitting statements of errors in the first account, only a qualitative agreement can be claimed between theory and experiment; science would not progress very far without being quantitative. Note that by replacing the "plastic rule technique" for fitting a straight line to linear data by the least squares method described later in this book, you also obtain an estimate of error on both the slope and the intercept. Not a bad return for a bit of computing.

You may have difficulty persuading yourself that the straight line in Figure 2.8 is a good fit to the points. This is because the top two points have twice the error of the other four and so have only a quarter of the weight. Judgement by eye, with which you are probably familiar, is not so obvious in such cases, but by all means check my working if in doubt; that would be good science.

Combining the equations quoted earlier gives

$$\frac{hc}{e\lambda} = \frac{A}{e} + V_{max} \qquad (4)$$

Thus a graph of V_{max} on the y axis against $1/\lambda$ on the x axis should yield a straight line, with a slope of hc/e and an intercept on y of $-A/e$.

In spite of all attempts to prevent light reaching the anode, some electrons were emitted from it, yielding a negative current that was evident when the voltage was large enough. This made it more difficult to decide the value of V to choose as V_{max}, i.e. when the cathode current, I, was just zero. The simple view taken here was to choose the value of V when the current was zero. This gives the data shown in Table 2.4, to which is added the weighted least squares fit analysis, shown graphically in Figure 2.8.

Table 2.4 V_{max} for different wavelengths and the weighted least squares fit analysis.

λ(nm)	$1/\lambda$(nm^{-1})	V_{max}(V)	dV_{max}(V)
651.6	1.535×10^{-3}	0.14	0.01
600.9	1.664×10^{-3}	0.36	0.01
550.1	1.818×10^{-3}	0.57	0.01
499.4	2.002×10^{-3}	0.82	0.01
449.8	2.223×10^{-3}	1.04	0.02
401.2	2.493×10^{-3}	1.32	0.02

Weighted least squares fit:

Slope $= (1.280 \pm 0.018) \times 10^{3}$ V nm
$= (1.280 \pm 0.018) \times 10^{-6}$ V m
Intercept $= -1.781 \pm 0.034$ V
CG point $= (1.822, 0.551)$

Correlation coefficient $r = 0.996$

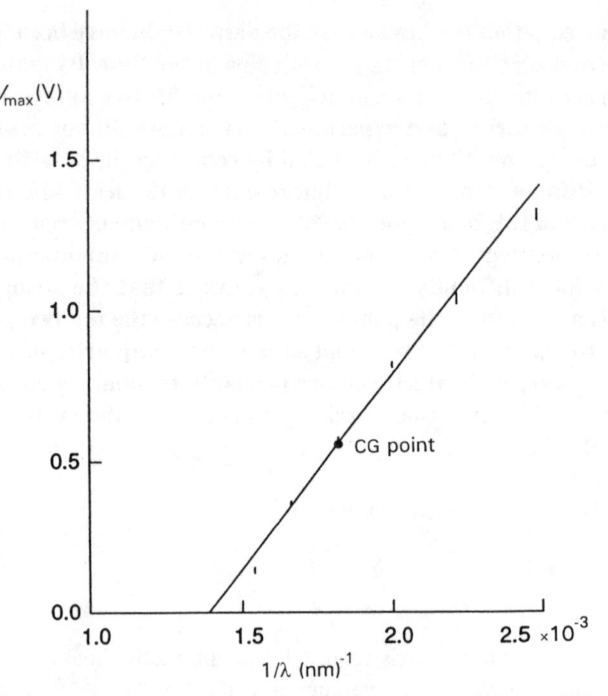

Figure 2.8 V_{max} versus $1/\lambda$: weighted least squares fit (CG, centre of gravity).

$$\begin{aligned} \text{Theoretical value of slope} &= hc/e = 1.240 \times 10^{-6}\ \text{V m} \\ \text{Experimental value} &= (1.280 \pm 0.018) \times 10^{-6}\ \text{V m} \\ \text{Difference} &= 0.040 \pm 0.018 \times 10^{-6}\ \text{V m} \end{aligned}$$

Since the difference is only a little more than 2 errors in magnitude the agreement between theory and experiment is reasonable.

$$\begin{aligned} \text{Intercept on } y \text{ axis} &= -A/e \\ &= -1.781 \pm 0.034\ \text{V} \\ \text{Therefore} \qquad A &= 1.781 \pm 0.034\ \text{eV} \end{aligned}$$

The unit of A is not SI, but is the one in common usage. This value cannot be checked since it depends on the composition and structure of the photocathode surface, which is not known. However, it is the right order of magnitude for surfaces used in photodiodes.

Discussion

This is another section that has to be tailored carefully to your reader. It is not a bad idea to suggest changes that you would have made had more time been available, and that the next student could adopt with profit. Any

scientist who thinks to have said the last word on a topic is doomed to disappointment. A friendly give-and-take between students in a class can work wonders for the progress of all of them.

The choice of V_{max} is probably the most contentious decision made in the photoelectric experiment, and suggesting the next student considers other possibilities could be helpful to both of you.

Perhaps the least satisfactory aspect of this experiment lies in the criterion used to choose the value of V_{max} from the measured data. A simple approach was made here, and though it yielded results that were acceptable, it would be worthwhile investigating other possibilities.

Another improvement may be achieved by concentrating only on those values of the current close to zero, as there is little useful information at larger values. Smaller increments of voltage, of say 0.01 V, might also be worth making.

Conclusion

This should be a simple statement answering the questions posed in your aims stated at the beginning of the report. If you think of these two sections as two parts of the same concept you will not go far wrong.

As before, errors and units were omitted from the first version to try to persuade you of their importance.

(a) Within the errors of measurement, Figure 2.8 is a straight line graph, thus verifying Einstein's photoelectric equation.
(b) The slope has a value $1.280 \pm 0.018 \times 10^{-6}$ V m, which compares satisfactorily with the theoretical value (hc/e) of 1.240×10^{-6} V m. The work function, A, of the photocathode surface is found to be 1.781 ± 0.034 eV. This value cannot be checked from standard tables, but is the right order of magnitude.

CHAPTER 3

Design of experiments

Introduction

By "experimental design" I mean performing all the steps necessary for the successful completion of an experiment. You may start with only a vague idea, but should aim to finish with a scientific report so that others also interested in the subject can evaluate your work, repeat it, or use it for further study.

It is rare for students to conduct an experiment all the way from beginning to end because of the long time required. So, although some of the steps will be familiar to you, others will be new. Even if you rarely apply the full procedure it is useful to know all the stages so that your own efforts are placed in a proper context.

Experimental science can be very intolerant of mistakes. You might perform 99% of a procedure correctly, but the 1% you get wrong may destroy all the good work you have done. The ability to recognize the sensitive parts of an experiment, such as those susceptible to systematic error, will come with experience, so do not be disheartened by mistakes made while developing your skills. The more training you have in performing experiments, and thinking about them, the faster will your expertise grow. Practice may not make perfect but it will certainly lead to steady improvement.

You must be prepared for results not anticipated during the early stages of the design, since clairvoyance is not a necessary characteristic for scientists. Nor must arrogance be part of your armour, since it is most unlikely that you will say the last word on any issue in science.

One question that we will discuss later is the efficient use of time in performing experiments. These often take a long time to complete so that it is worth developing an attitude that seeks the quickest route to a particular end, rather than just drifting along in a haphazard manner.

Steps along the way

We will start with a list of the necessary steps, discuss each in general terms, and then apply the ideas to particular experiments:

1. Choose a subject suitable for an experiment.
2. Discover what is known about it.
3. Have you the necessary facilities?
4. Design the experiment.
5. Make preliminary measurements.
6. Analyze the preliminary results.
7. Redesign the experiment if necessary.
8. Perform the experiment.
9. Write a scientific report.

General discussion

1. Choice of subject

In industry a suitable subject might arise from a problem encountered in a manufacturing process, such as the need to make a product more reliable or cheaper. University research may be more concerned to ask fundamental questions about the way the universe functions. A teacher in a classroom selects experiments based on the ideas, in physics or experimentation, at a level suitable for the students concerned, or sets tasks to practise a particular skill.

Whatever the subject, from whatever source, you must remember that a lot of time and effort will be needed before a satisfactory solution is achieved. You should be quite sure, therefore, that you are willing to make the necessary investments before continuing. It is also important to define the problem clearly, even at this early stage. Of course the emphasis may change as you proceed to investigate it, but that is no excuse for lack of clarity at the outset.

2. What is known about the subject?

Some scientists feel the need to find out virtually all that is known about a subject before designing their own experiment. This has the danger that reading a well written scientific report may encourage you to believe that there is no further problem to be investigated, though this is rarely the case. At best it will have answered the questions thought appropriate at the time the work was performed, and you should not need convincing that science moves forward, sometimes at an alarming pace.

Other scientists would agree with Sir Edward Bullard: "I think it is best to work for a bit at a subject before reading what other people have done. If you read other people's papers for several days on end you get into their way of thinking and may miss ideas that would otherwise have occurred to you." Keeping these extremes in mind you should develop a method which suits your temperament.

3. Have you the necessary facilities?

The facilities needed will vary enormously with the scale of the problem to be investigated. Any attempt to understand the interactions between elementary particles will probably require a large accelerator, tons of electronics, including very fast computers, and a large team of physicists, engineers and technicians.

As a chief examiner for A-level physics I was aware of the need to set problems around equipment likely to be available in all schools entering candidates for the exam. One's popularity with teachers was always likely to rise if experiments were devised around equipment which would be of value after the exam, though I must confess to failing to base an experiment around squash balls or malt whisky.

Occasionally the equipment needed for a particular investigation is not available because no one has invented it yet. Then the ability of some scientists to design and construct a new piece of equipment may come to the fore. Vacuum pumps and detectors of nuclear radiation are examples of experimental tools which have been developed over long periods of time. In more recent times the laser was at first described as a solution in search of a problem, but not for long.

Naturally the universal facilities are time and money, but who has ever had enough of either?

4. Design the experiment

This means doing all the preliminary work needed to ensure a high probability of success in the experiment: choice of equipment, conditions of measurement, estimating the likely accuracy, control of unwanted variables, and so on.

The possibility of systematic errors should never be far from your thoughts at this stage. It may be necessary to make a few simple measurements to test the sensitivity of your apparatus to a particular experimental variable before choosing it for your final experiment. You do not want to discover during the final measurements that part of your apparatus requires a temperature stability of ±0.01 °C, say, when you have only allowed for ±0.1 °C. On the other hand you do not want to spend time and money providing elaborate conditions that are not necessary. The well designed experiment *does what is necessary, but no more*. You can always improve the experiment later, as Millikan did when measuring the charge on the electron.

5. Make preliminary measurements

This will enable you to get a "feel" for the apparatus. Is it sensitive enough? Is it easy to adjust? Can the scales be read correctly? And so on. You would be a remarkable person if you could use an unfamiliar piece of apparatus as effectively the first time as you will with more practice.

Enough data should be taken to enable you to make a reasonable estimate of the random error likely to be achieved in the final experiment. If this is unacceptably large then redesigning is needed.

In many experiments measurement of one variable makes a larger contribution to the overall error than the others. The preliminary measurements should identify this variable. You may then decide to reduce its contribution to the error in some way: changing the apparatus used for its measurement or increasing the number of measurements. If this means less time being spent on those measurements that are already precise then that represents a good use of time.

6. Analyze the preliminary measurements

This section might have been included under the previous one, since there is little point in making preliminary measurements if you do not analyze them. But part of your design strategy will have included the method of analysis of the data, and it is worth considering at this stage whether your original choice was the best one. Chapter 5 discusses some of the methods available.

7. Redesign the experiment, if necessary

After steps 5 and 6 you should be able to decide whether the aims of the experiment are likely to be reached when the final measurements are made. If there is any doubt, then a redesign of the experiment must be contemplated. This could involve anything from a simple change of technique to a complete overhaul of the original idea, and even its abandonment. If you are really lucky, and particularly alert, you may make an unexpected discovery such as the modification of Stokes's law for small particles found by Millikan while measuring the charge on the electron.

8. Perform the experiment "proper"

This is the part of experimental design that students are most familiar with, as the earlier steps are usually carried out by your teacher to ensure that reasonable results can be obtained in a limited time. Remember to compare your results with those taken by other people so that systematic errors are brought to light, if present to a significant extent. Ignorant people may call this "cribbing", but if done properly it is good science.

9. Write a scientific report

Free communication is the lifeblood of science, so you must take this step seriously. It is sufficiently important to warrant a separate chapter in this book (see Ch. 2).

National security or commercial profit may be used as an argument for limiting the communication of scientific information, and your education in science might consider such topics. At a conference I attended some years ago a scientist was happy to listen to work performed in university laboratories but was unwilling to discuss his own work, which he considered to be commercially sensitive. He was not very popular.

Good use of time in experiments

There are occasions when you are short of time to conduct an experiment as thoroughly as you would wish. Students may be in this position during a practical examination, or astronomers investigating a fleeting phenomenon such as the passage of a comet or an eclipse. A good plan of campaign is essential if you are to get the best out of such situations.

Sometimes you can arrange for two things to continue at the same time: electronic equipment warming up while you arrange the rest of the apparatus; a printer producing copies of computer files while you analyze experimental data.

Even if you do have plenty of time for the experiment it is sensible to apportion your effort where it will produce best results. i.e. smaller random or systematic errors. Conversely, what is the point of making a large effort in a part of the experiment that has little influence on the final result? You can only claim to understand an experiment properly if you appreciate both its strong and weak points.

Example 3.1 – measuring the density of paper

Assume you start with a single sheet of A4 paper, 30 cm rule, micrometer screw gauge, and top pan balance. Using the good practice discussed earlier we decide to perform a preliminary experiment.

As a rough estimate of the random error in the various measurements let us assume a value of 1 in the last figure quoted in each case. Of course we can get a "proper" value by repeating measurements and calculating the standard error on the mean, but we only need a rough guide at the moment in order that we can calculate the relative contributions to the error on the density. Let us say that the four measurements give the following values:

	Length:	l	$= 29.7 \pm 0.1$ cm
	Width:	w	$= 21.0 \pm 0.1$ cm
	Thickness:	t	$= 0.011 \pm 0.001$ cm
	Mass:	m	$= 5.19 \pm 0.01$ g
Thus	Density:	ρ	$= m / lwt$
			$= 0.756\,483$ g cm^{-3}

You may think it silly to quote six figures for the density when no contributory measurement is quoted to better than three. You are right of course, but I want to emphasize that it is better to quote too many figures than too few, since it is easier to drop figures later than to regenerate them. We can only be sure how many figures to quote when we have calculated the error on the density, σ_ρ. To do this we use the formula discussed in Chapter 5, and assume there is no correlation between the four measurements:

$$
\begin{aligned}
(\sigma_\rho/\rho)^2 &= (\sigma_l/l)^2 + (\sigma_w/w)^2 + (\sigma_t/t)^2 + (\sigma_m/m)^2 \\
&= (0.1/29.7)^2 + (0.1/21.0)^2 + (0.001/0.011)^2 + (0.01/5.19)^2 \\
&= 1.13 \times 10^{-5} + 2.27 \times 10^{-5} + 8.26 \times 10^{-3} + 3.71 \times 10^{-6} \\
&= 8.30 \times 10^{-3}
\end{aligned}
$$

therefore $\quad \sigma_\rho = 0.069\,\mathrm{g\,cm^{-3}}$

so $$\rho = 0.756 \pm 0.069\,\mathrm{g\,cm^{-3}} \qquad (5)$$

Comments

(a) We have achieved a random error of about 10%. If this is sufficient for our purpose there is no need to do any more work, and the preliminary measurements have become the final ones.

(b) Values for the density of cellulose are quoted in the *CRC handbook of physics and chemistry* (a standard reference text) to be in the range 0.7– 1.15 g cm^{-3}, and our measurement is within this, albeit broad, band.

(c) The predominant contribution to the error on density arises from the measurement of thickness, t. This is therefore the weak link, and its measurement must be improved if we want a better accuracy for density. *We must not waste time improving the accuracy of the other three measurements.*

(d) If an error of less than 10% is required there are various ways of improving the accuracy of measurement of thickness:
   (i) σ_t could be reduced by using a more sensitive thickness gauge, perhaps based on the wavelength of light.
   (ii) A little improvement could be obtained by interpolation between the scale readings of the micrometer screw gauge, but this would demand great consistency in the force used to close the jaws on the paper. Repeating readings does not reduce the error much, though it would be useful to check that the thickness was uniform over the whole area.
   (iii) The simplest approach might be to increase t by folding the paper a number of times. Twenty thicknesses of paper are enough to reduce the error on t to less than that for w, the next biggest contributor. If we used 20 thicknesses, then the new weak spot would be the measurement of the width of the paper, w, and we would have to consider how to reduce the error on that if further improvement was required.

The approach outlined above enables us to steadily improve the precision of measurement in an efficient manner by identifying the weakest spot in the experiment and putting effort into improving it. Although it gets progressively harder to keep reducing the random error, we can often do so with sufficient effort. But it is silly to spend a lot of time and effort reducing the random error only for it to be swamped by a larger systematic error.

The way to search for systematic errors is to perform more measurements, preferably using a different approach. Unfortunately, the alternative technique for measuring density, Archimedes' method, requires a liquid that does not wet paper. Even if you do not use a different method it makes sense to ask other people to repeat the measurements in case your handling of the equipment has introduced a systematic effect. If a class of students perform the experiment, one can check for the consistency of density for a given batch of paper, or between batches.

Example 3.2 – counting weak radioactive sources

The strength of a radioactive source cannot be found from a single measurement because some of the counts recorded will arise from background radiation rather than from the source. You have to take two measurements:

(a) source + background, giving n_{s+b} counts in time t_{s+b};
(b) background only, giving n_b counts in time t_b.

The source strength is then deduced from the difference between these two measurements.

Given a fixed duration for measurement, how should we apportion the total time between these two? One can see that the optimum division of time will vary with the strength of the source with the following reasoning. If the source is very strong compared with the background we do not need to know much about the latter because its effect is small. In other words we spend most of the time counting the source. At the other extreme, when the source is very weak, the counting rates in the two measurements are roughly equal, and there is little to distinguish one from the other. One would thus spend equal times on the two measurements. Let us now put this qualitative idea onto a quantitative footing.

We will work in terms of the counting rates (r, in counts per second) because these represent the strengths of source and background. Thus

	r_{s+b}	$= n_{s+b} / t_{s+b}$	Source + background
and	r_b	$= n_b / t_b$	Background only
so	r_s	$= r_{s+b} - r_b$	Source only

As with any experiment there will be random errors in measuring r_{s+b} and r_b. These will propagate into an error in r_s, which I will call σ_{r_s}. The most accurate measurement will be achieved when we divide our time between the two measurements in such a way as to produce a minimum for σ_{r_s}.

Using the rule for the propagation of errors of a difference from Chapter 5:

$$\sigma_{r_s}^2 = \sigma_{r_{s+b}}^2 + \sigma_{r_b}^2 \tag{6}$$

but radioactive decay obeys Poisson statistics, for which the standard deviation on n counts is simply $n^{1/2}$, so, dividing by the appropriate value of t to get the counting rates:

$$\sigma_{r_{s+b}} = n_{s+b}^{1/2} / t_{s+b}$$

and

$$\sigma_{r_b} = n_b^{1/2} / t_b$$

therefore

$$\sigma_{r_s}^2 = n_{s+b} / t_{s+b}^2 + n_b / t_b^2$$

$$= r_{s+b} / t_{s+b} + r_b / t_b \tag{7}$$

The total time, T, used for both measurements is the sum of the individual times:

$$T = t_{s+b} + t_b \tag{8}$$

Eliminating t_b from equations 7 and 8 in order to calculate the fraction of time used in counting the source gives

$$\sigma^2_{r_s} = r_{s+b}/t_{s+b} + r_b/(T - t_{s+b}) \tag{9}$$

The condition for a minimum in $\sigma^2_{r_s}$ is that the differential coefficient is zero, i.e.

$$d\sigma^2_{r_s} / dt_{s+b} = 0$$

therefore $$-r_{s+b} / t^2_{s+b} - r_b / (T - t_{s+b})^2 \times (-1) = 0$$

and re-arranging gives $$t_{s+b} / T = 1 / \left[1 + (r_b / r_{s+b})^{1/2}\right] \tag{10}$$

This seems to require the *answer* to the measurements (r_b/r_{s+b}) before we can decide the best conditions for making them! The apparent illogicality can be avoided with a quick preliminary measurement, lasting perhaps a few minutes, to find an approximate value for r_b/r_{s+b} before starting the final measurements lasting a few hours, say. We will investigate the complete range of the variables, but three of the values are worth special comment:

r_{s+b}/r_b	Optimum t_{s+b}/T	Comments
1	0.5	Very weak source
2	0.586	Source and background equal
5	0.691	
10	0.760	
20	0.817	
50	0.876	
100	0.909	
200	0.934	
500	0.957	
1000	0.969	
5000	0.986	
Infinite	1	Very strong source

Note that the first and last items in the table correspond to the special cases discussed qualitatively earlier. It is gratifying to finish with a solution that "feels" correct, though there is no way we could have arrived at intermediate values without the analysis.

These results would be of particular value if you were employed in the regular testing of items of food for radioactive content, a topic which took special importance in Europe after the accident at Chernobyl.

To summarize, there is always an optimum way of using your time in performing an experiment. This is based on the need to reduce the error on the final value to a minimum, given the constraints of time and apparatus available. A lengthy analysis may be needed to arrive at the optimum conditions, though we might get a rough idea by using our knowledge of what happens in extreme cases.

Example 3.3 – charge on the electron

(*Note on the units of electronic charge*: Physicists above a certain age will tell you, with a faraway look in their eyes, of the times when life was so complicated that charge had to be measured in two different units – electrostatic (esu) and electromagnetic (emu). You have been born into a simpler world which recognizes that the static and magnetic properties of charge are just two aspects of a single cause, so that we now use a single unit, the coulomb. The charge on the electron is 1.602×10^{-19} C, or 4.803×10^{-10} esu in the units quoted by Millikan.)

The classic experiment is that first performed by R. A. Millikan, and after a short introduction the steps in experimental design will be applied to his work.

Briefly, the experiment consists of spraying very small droplets of oil between the plates of a parallel plate capacitor and selecting a drop which has collected a charge by friction. The balance of the gravitational pull downwards and the viscous drag upwards results in a steady speed of fall, v_d. Before reaching the bottom plate an electric field is applied in a direction such as to move the drop at a constant speed, v_u, upwards. The drops are so small that they have to be viewed through a microscope, and with the careful application and removal of the electric field may be observed many times in transitions up and down.

If the drop has a radius r, density ρ, moving through a medium of density σ, and coefficient of viscosity η, the resultant force downwards (F_d) is the weight of the drop (F_w) less the upthrust (F_u) of the medium, air, and the viscous drag (F_v) through it:

$$\begin{aligned} F_d &= F_w - F_u - F_v \\ &= 4\pi r^3(\rho - \sigma)g/3 - 6\pi\eta r v \end{aligned} \qquad (11)$$

For a drop of suitable radius this force can become zero so that the drop descends with constant speed v_d.

The second measurement is to apply a constant electric field, generated by a potential difference V across parallel plates separated by a distance d, which produces a force (F_e) on the charge q of the drop to yield a resultant force (F_{up}) upwards (the weight and upthrust act in the same directions as before, but the viscous drag is now downwards, opposing the motion as always):

$$F_{up} = Vq/d - 4\pi r^3(\rho - \sigma)g/3 - 6\pi\eta rv \qquad (12)$$

Again this can become zero, for the same drop, provided V has an appropriate magnitude, yielding a steady upward speed, v_u.

The most difficult variable to measure in these two equations is the radius of the drop, r, so we eliminate it from Equations 11 and 12, and after a little algebra, obtain the following expression for the charge, q, on the drop:

$$q = [6\pi\eta d / V][v_d - v_u][9\eta v_d/2(\rho - \sigma)g]^{1/2} \qquad (13)$$

Let us now look at some of the design considerations.

1. Suitability of the subject

The suitability of scientific research can best be judged in its historical context. What is at one time impossible, later becomes difficult, and later still routine. During the nineteenth century two theories competed to explain the source of electrical phenomena. One considered charges sitting on bodies, and the other as strains in the medium. Millikan argued that it is one thing to suppose that an electrical charge *produces* a strain in the surrounding medium, but quite another that it *is* that strain, just as a person standing on a bridge produces a strain in it, but cannot be described as merely a strain in the bridge.

The practical difference between the two points of view is that the former encourages us to look for attributes of the person other than the effect on the bridge. Similarly, we should look for properties of electrical charge other than the strain that it produces in the medium. The attribute looked for by Millikan was the fundamental unit of charge. Did it exist and could it be measured?

2. What was known about the subject?

Millikan was not alone in asking such questions, and measurements on charged clouds had already been made:

Year	Investigator	Charge (C)
1897	J. S. E. Townsend	1×10^{-19}
1898	J. J. Thomson	2.2×10^{-19}
1903	J. J. Thomson	1.1×10^{-19}
1903	H. A. Wilson	1.0×10^{-19}
1908	R. A. Millikan	1.35×10^{-19}

The range in values of the unit of charge, tabled above, was unsatisfactorily large, perhaps inevitably so due to the use of clouds, with the inevitable averaging over a wide range of drop sizes. Perhaps even the concept of a fundamental unit of charge was invalid?

Millikan then made the decisive change in experimental design by making all the necessary measurements on the *same* charged drop of oil, rather than forming an *average* over a whole cloud. He then not only had a problem worthy of his considerable talent but an idea that could be developed into a suitable solution.

3. Had he the necessary facilities?

Hindsight makes it clear that Millikan did have the necessary facilities. The beautiful simplicity of the apparatus he designed might make the experiment appear simple to us now, until you try to repeat it. Then you will marvel at the sheer persistence required to produce so many meaningful results.

4. Design the experiment

Three elements of the design are worth special mention.

(a) The typical speed of an oil drop, 0.05 cm s^{-1}, made it particularly important to reduce convection currents to a minimum. This was achieved by surrounding the apparatus with an oil bath maintained at a temperature stable to 0.02°C during a measurement.

(b) The drops of oil had to be dry and free from dust if they were to be consistently affected by the gravitational and electrical fields applied to them.

(c) Measurements were made on drops with various radii and number of fundamental charges held. Thus, any possible variation of density or drag with radius of the drop could be investigated. This paid dividends when Millikan discovered the need for a modification to Stokes's law for the viscous drag on very small spheres.

5. Make preliminary measurements

The results of Millikan's preliminary experiments were not as expected: the measured elementary charge should have been independent of the radius of the oil drop to which it was attached. Unfortunately, a graph of charge against velocity of fall under gravity showed discrepancies at low speeds (i.e. small radii). Millikan correctly assumed that Stokes's law was not valid for the smallest drops observed, those with radii about 0.0002 cm. For such small radii, comparable to the distances between molecules, he postulated that the medium through which the drops fell no longer appeared homogeneous. Not only did this understanding explain his results, but suggested the line of approach by which they could be corrected.

6. Analyze the preliminary measurements

The preliminary measurements had thrown sufficient doubt on the validity of Stokes's law that further investigation was needed. This sought to justify the five assumptions made in the theoretical derivation of Stokes's law:

(a) Inhomogeneities in the medium are small in comparison with the size of the sphere.
(b) The sphere falls as it would in a medium of unlimited extent.
(c) The sphere is smooth and rigid.
(d) There is no slipping of the medium over the surface of the sphere.
(e) The velocity of the sphere is so small that the resistance to motion is due to the viscosity of the medium and not due to the inertia of some of the medium being pushed along by the sphere.

Some ingenious measurements involving spheres as small as 0.002 cm radius settled these doubts and enabled Millikan to analyze his data accordingly. A simple multiplying factor of $(1 + b/pr)$ had to be made to the steady speed of fall predicted by Equation 11; b being an experimental constant and p the pressure of the gas in which the drop descended.

7. Redesign the experiment

The experiment was redesigned in the sense that the final value of e was based on observations made under conditions close to the optimum. The size of drops preferred was a compromise between being small enough to allow accurate timing, but not so small as to give a large correction for the deviation from Stokes's law. Then one would not need to know the correction factor very accurately.

The value quoted was 4.774×10^{-10} esu. This is equivalent to 1.592×10^{-19} C, a little lower than the value currently accepted.

Realizing the fundamental importance of his measurement of the basic charge, *e*, Millikan was determined to do even better. The apparatus was thus redesigned. The factors discussed below are meant to show how much fine detail has to be considered in a careful experiment. In 1914 a new electrode system was built with the surfaces optically flat to within 2 wavelengths of sodium light. They were separated by three spacers, each 14.9174 mm thick, with optically plane-parallel surfaces. The potential difference was compared with that from a Weston standard cell, and times were recorded with a chronograph that printed to an accuracy of 1/100th of a second.

8. Perform the final experiment

The work was concluded in August 1916, having occupied the better part of two years.

9. Write a scientific report

Reference: R. A. Millikan, *Philosophical Magazine* **34**,1, 1917. There had been other reports on earlier parts of the work but this is the final one.

Example 3.4 – the Mpemba effect

This is a very different example from the pioneering work of Millikan, but is included to illustrate the importance of a questioning attitude to science and the complexity of an apparently simple phenomenon. The problem is posed thus: given two containers of water, alike in every respect except that one is at a higher temperature than the other, which will freeze the faster if placed into the same freezer? Common sense tells us the answer is so obvious that we are justified in doubting the sanity of the questioner. We may well have some sympathy for Erasto Mpemba's teacher who showed little respect for the pupil's suggestion that the initially warmer liquid froze the faster. After all it would not be the first time a student had tried to divert a lesson into a more agreeable direction!

1. Suitability of the subject

It is the unexpected nature of the topic that makes this a suitable subject for experimental investigation. There seems to be no new fundamental physics involved, and it is unlikely to lead to commercial dominance of the ice-cream market for the manufacturer who reads the scientific literature. However it is likely to deepen our understanding of the application of physical principles to practical situations.

2. What was known about the subject?

Initially, knowledge of the subject was limited to an observation best described in Mpemba's own words:

> In 1963, when I was in form 3 in Magamba Secondary School, Tanzania, I used to make ice-cream. The boys at the school do this by boiling milk, mixing it with sugar and putting it into the freezing chamber in the refrigerator, after it has first cooled nearly to room temperature. A lot of boys make it and there is a rush to get space in the refrigerator. One day after buying milk from the local women, I started boiling it. Another boy, who had bought some milk for making ice-cream, ran to the refrigerator when he saw me boiling up milk and quickly mixed his milk with sugar and poured it into the ice tray without boiling it, so that he may not miss his chance. Knowing that if I waited for the boiled milk to cool before placing it in the refrigerator I would lose the last available ice-tray, I decided to risk ruin to the refrigerator on that day by putting hot milk into it. The other boy and I went back an hour and a half later and found that my tray of milk had frozen into ice-cream while his was still only a thick liquid, not yet frozen.

3. Have you the necessary facilities?

The facilities required to investigate this problem should be available in any laboratory: freezer, thermometers, containers for water, etc. Some means of reading temperatures automatically and perhaps plotting them on a pen recorder or computer screen would also be an advantage. If this is done without periodically opening the freezer door then better temperature stability will ensue.

4. Design the experiment

The first step in the experimental design might be to repeat the basic experiment under better control, by ensuring equal amounts (volume or mass?) of liquid are used in the two samples. It would also be necessary to note the positions used for each sample since it is likely that some parts of the freezer will be colder than others. The experiment could then be repeated with the positions of the two samples interchanged.

Having been satisfied that the phenomenon is true, one should then consider the experimental variables to be used in later measurements. These could include the surface areas exposed to the air in the refrigerator, the initial temperature difference between the samples, the insulation properties of the containers, etc. If for no other reason than one has to decide which experiment to perform first, it is worth trying to think about theoretical reasons for the phenomenon. It is not good practice to take measurements without at least a rough theory for guidance. Further design considerations are left as an exercise for the reader.

5. Make preliminary measurements

The preliminary experiments will attempt to decide which factors are important in bringing about the effect, as well as those variables which are not important. If one takes the simple approach, and investigates a single variable at a time, this will necessitate many experiments.

6. Analysis of preliminary results

This will result in lists of important and negligible factors, with magnitudes of their effects.

7. Redesign the experiment

From the many preliminary measurements this will seek to establish the crucial tests for understanding the phenomenon.

8. Perform the experiment proper

This might be a doddle after the large number of preliminary measurements! At least there is no harm in hoping so.

9. Write a scientific report

Usually it is a good idea to write about success rather than failure, so you may well decide to omit some of the experiments that turned out to be blind alleys. You may care to read some of the scientific papers written on this subject, starting with the first: E. B. Mpemba & D. G. Osborne, *Physics Education*, **4**, 172, 1969.

CHAPTER 4

Measuring the experimental variables

Introduction – a general experimental system

Many, if not all, experiments consist of three elements in which there is an explicit or implied balance represented by the equation

$$\text{Input} + \text{System} = \text{Output}$$

Usually you have information about two of these components and will attempt to make deductions about the third. We will briefly consider three examples in which the unknown being investigated, shown in bold print, shifts from input to system to output:

Experiment	Input	System	Output
1	**Starlight**	Spectrometer	Wavelengths
2	X rays	**Crystal structure**	Angles
3	Spectral lamps	Spectrometer	**Angles**

Experiment 1 uses the radiation from a star as the input to a telescope with a spectrometer attached so that the wavelengths of the light can be measured. By comparing these wavelengths with those produced by similar sources on Earth one calculates the red shift in the light emitted by the star, from which the rate of expansion of the universe is deduced.

In experiment 2 one measures the angles at which X rays are diffracted by a crystal to find how the atoms are arranged inside. The discovery of the double helix structure of DNA is perhaps the most famous example of such measurements.

Calibration of a spectrometer is achieved in experiment 3 by measuring the angles at which radiation of fixed wavelengths emerge from the instrument. Then one can use the spectrometer to deduce the wavelengths of unknown spectral lines from the measured angles.

Conservation of energy is another familiar example of the balance between input and output, though confidence in its validity was sorely tested in the 1930s when an explanation was sought for the continuous energy spectra found in the β decay of radioactive nuclei. While Bohr was prepared to consider the violation of the principle of conservation of energy, Pauli preferred to postulate the existence of a new particle, the neutrino, a view which has prevailed. Even he seems to have delayed publishing his idea, and it was not until 1956 that the particle was discovered experimentally by Cowan and Rienes.

Equipment is needed to make measurements in each of the three sections of an experiment, and the characteristics of that equipment and the skill with which it is used is an important topic in experimental physics. These characteristics will first be discussed in general and then applied to specific examples.

Characteristics of equipment

The perfect piece of equipment is accurate, easy to use, and comes free in a cornflake packet.

Real equipment is more complex, and which of its characteristics are most important will depend on the particular circumstances, so the following list is simply alphabetical:

(a) accuracy;
(b) complexity;
(c) cost;
(d) ease of use;
(e) longevity;
(f) output;
(g) range;
(h) reliability;
(i) sensitivity.

Accuracy

This is a mixture of two concepts: precision and calibration.

Equipment is precise if it gives the same output for a given input over the required period of time. In other words it gives reproducible results. But there may be systematic errors in every reading that make them all false. In order that the equipment be not only precise but accurate we also need to know that it is properly calibrated. If you use a micrometer screw gauge to measure the thickness of some object, you may obtain a precise result if the instrument has a zero error, but it will not be an accurate measurement unless you remove or allow for the error. There is still the possibility that the scale itself is not calibrated properly, but it is more difficult to investigate and correct such an error, so we will leave that to the competence of the manufacturer.

Complexity

The time-honoured rule in life is that when all else fails, read the instructions! Unless, that is, they are a poor translation from the Japanese, in which case it is quicker to ask someone who is familiar with the equipment. This was the situation I was in a number of years ago when trying to make a computer and printer communicate sensibly. You could take this as a warning to try and understand the needs of the recipient when writing something for another person to read.

Manufacturers have the opportunity to standardize certain common features of their equipment, such as the symbols used to represent certain functions, but often fail to do so unless prompted by international agreements. The "record" button on my video recorder is identified by "REC" printed in red, while my son's has a ring of dots. Although standardization too early in the development of equipment may stifle progress, no one is going to produce a video recorder without a record button or make dramatic changes to this function.

Perhaps the greatest example of the failure to standardize is in computer software. The cynic in me wonders if all self-respecting computer scientists have to develop their own language. It may not matter much for applications requiring only a few lines of code, but what about safety-sensitive software used to control a nuclear power station or an air traffic control system. Try changing one of those programs to make use of up-to-date hardware.

Once I was supervising a very good student who was making measurements of the viscosity of a gas by timing its flow through a narrow tube. The results appeared very strange until it became apparent that she had read the stopwatch as 110 s, say, instead of 1 min 10 s. Although the units were correctly displayed on the watch they were not at all clear until one looked hard.

Cost

No one wants to pay more than is necessary for a piece of equipment, but getting value for money is not just about buying the cheapest item on the market. After all, I subscribe to various magazines to profit from other people's experience and not rely on gaining insight by hindsight. The skill is in buying what you want without paying extra for functions not needed.

Correct timing of a purchase is also important if the technology is changing rapidly. I took a number of years before buying my first computer, and was persuaded to do so by its ability to interface with sensors of various types. Ten years or so later I awaited the arrival of its successor before upgrading to my present beautiful machine. Two good choices out of two, so far, is satisfactory but the next change will have to be considered equally carefully.

It always surprises me that many university departments do not allow for depreciation of equipment from year to year, based on its cost and likely lifetime, but seem to be taken by surprise each time an expensive set of apparatus needs upgrading. Mind you, if inflation rates are based on the contents of an everyday shopping basket, rather than that of specialized equipment, it is not surprising that grants for science fall short of needs.

Ease of use

The world seems to be divided into two groups of people: those who can and those who cannot programme a video recorder. Of those who can, only a minority are foolish enough to attempt to programme one of a different make: people are so intolerant when you record a much repeated film instead of their favourite football team!

This book is being written on a word processor cum desktop publisher. Like all substantial software it is "easy to use" only after a considerable time has been invested in training and practice. My publisher does not have a similar system, and since both of us seem content with what we have, it was necessary to check that disks of information could be transferred between the two. As with other issues in experimental physics you must invest whatever resources are necessary to be "at ease" with the equipment, including software; too little is insufficient, too much is wasteful of effort.

Cameras from earlier times required you to set lens aperture, shutter speed and focusing distance, so that the bird could have flown or the sun hide its face before you were ready to take a photograph. Modern versions incorporate so much automation that it is almost simply a matter of "point and press". Would that computer manufacturers copied this trend to simplicity.

Longevity

If I can stay around for a long time why cannot my equipment? Since time is the variable under consideration it would be appropriate to discuss this topic historically.

In the days when equipment was mainly mechanical it would often last a long time provided materials of good quality were used and the construction was sound. Maintenance was often simply a case of applying oil and grease at appropriate places and at regular intervals. Working steam engines pay tribute to the longevity achievable.

For many years electronic circuits had to use vacuum valves, which used thermionic emission from a hot cathode as a source of electrons. In bulk these generated large quantities of heat, which made them welcome companions in cold, winter laboratories, but which could lead to unreliability and breakdown as components became too hot for comfort. If the valves were distributed in space to reduce the energy density one could be left with a small desk in the corner of a room otherwise occupied by a glowing beast.

The point contact transistor was invented in 1947 by Bardeen and Brattain, closely followed by the junction transistor in 1950 by Shockley. These generated less heat than vacuum valves for a given effect, but as with any new discovery needed development time to reach their inherent level of reliability.

Integrated circuits carried this principle one stage further, though they created the temptation to pack circuitry into as small a space as possible. After all, the ultimate frequency of operation of a circuit is determined by the speed of light, so that reducing the separation of components is a clear way of increasing speeds. Fans are then needed to keep the components cool, and one is purring away in my computer at the moment. Such apparatus is potentially longlived, and its lifetime is more likely to be determined by new needs rather than the reliability of the equipment.

Software has been around for less time, and perhaps its very flexibility makes the effective lifetime short, since it is subjected to updated versions arising from the need to remove bugs and provide facilities not thought of at the time of creation. It has also got to submit to changes in the hardware used in the computers on which it runs.

Output

In the bad/good old days of physics the output from many instruments took the form of a pointer, mechanical or optical, moving over a scale. A better class of instrument would incorporate a mirror behind the pointer so that parallax could be avoided by aligning the pointer and its image before reading its position on the scale. Scale markings could not be so tightly packed that confusion would arise, so that interpolation was expected if the pointer settled between two marks. Zero readings could either be subtracted from the final reading, or a screw might be provided to set the pointer over the zero mark with no input. Limitations in precision arose either from the finite length of the scale or thickness of the pointer. Damping of the movement of the pointer was normally set at its "critical" value so that the final position was reached as quickly as possible without overshoot or oscillation. The petrol indicator in my present car fails in this respect, as its movement is so heavily damped that it takes too long to reach its final position.

If the output is in the form of an electrical signal the following three types of display are possible. The first is in digital form, with the number of digits determining the precision achievable. If you are lucky, or spend enough money, a range of digits will be available so that you can match the precision of the meter with the stability of the signal being measured. The ideal is to have some variation in the final digit; fewer digits will give less precision than is justified, more will merely give random fluctuations in the later ones.

Secondly, a permanent record of an analogue signal can be achieved by using a pen recorder. This is a continuous roll of paper, moved by a drum turning at a fixed speed, with a pen making a mark at right angles to this motion at a position along the paper proportional to the signal. In effect one achieves a graph of voltage against time. A further development of the same idea is to make the progress of the paper proportional to a second input voltage, rather than time through the constant speed motor. This produces a piece of equipment called an X–Y plotter. Because of the common features between these devices it is hardly surprising that some pieces of equipment incorporate a switch to provide either function.

Finally, the next logical step is to put the signal into a computer, which can do all that a pen recorder or an X–Y plotter can do, with the extra facility of making calculations and statistical analyses. The danger then is that one is spoilt for choice in the method of displaying the results of an experiment, and you have to think carefully which is most appropriate.

Range

It is a rare piece of equipment that covers the whole range of a particular experimental variable. This is well illustrated by the International Temperature Scale of 1990, which spans more than 1300 degrees to satisfy all its users. This is achieved by specifying types of thermometer, precise conditions of use, and the fixed points at which they must be calibrated. Interpolation between the fixed points is then possible.

The total range of the scale is covered by thermometers based on the following thermal phenomena:

Range (K)	Phenomenon
0.65–5.0	Vapour pressure of ^{3}He and ^{4}He
3.0–24.5561	Helium gas thermometer
13.8033–961.78	Platinum resistance thermometer
>961.78	Planck's radiation law

There are 17 fixed points, starting with the vapour pressure of helium at 3 K and ending with the freezing point of copper at 1357.77 K. The thermodynamic unit of temperature, the kelvin (K), is defined as 1/273.16 of the temperature of the triple point of water (the triple point is the point at which there is thermal equilibrium between the three states of matter). This puts into perspective the range of 273–373 K covered by the type of thermometer with which you are most likely to be familiar, the mercury in glass thermometer.

What equipment can be built also depends on the properties of the materials available for its construction. The property which has the largest range of values is resistivity, with conductors such as copper at one end and insulators made from a range of plastics at the other. Semiconductors come somewhere between these extremes and are arguably more important than either in modern apparatus.

Reliability

The only thing worse than a piece of equipment that is unreliable is one that appears to be working correctly but gives wrong results. This problem is obviated by calibrating regularly, and if you can alternate calibration and measurement at suitable intervals you should also spot any periods of unreliability. The results obtained during this period can then be discarded if they cannot be corrected. An intermittent fault is a curse because its time of occurrence is unpredictable, though continuous monitoring of the signal, with a pen recorder for example, will show when it has occurred.

Many years ago, in the days of thermionic vacuum valves, I built a piece of equipment that would now be called a pulse height analyzer. This measured the amplitude of a voltage pulse produced when a nuclear particle entered a detector called a scintillation counter. The amplitude is a measure of the energy deposited in the counter by the particle, so that with sensible precautions one can measure the energy spectrum of particles emitted from the radioactive source, which can therefore be identified and its intensity measured. The circuits were designed to detect pulses at a maximum rate of 10^6 per second, but were unreliable at this frequency. By reducing the operating frequency to 8×10^5 per second, reliable performance could be achieved. This simply meant that measurements took a longer time to gain the required accuracy, and since there was a large element of automation in the experiment it merely increased the length of coffee breaks - sorry I mean time in the library! This was considered to be a better use of the equipment than going into great detail trying to find the cause of the problem.

Sensitivity

This is the smallest signal that the equipment will detect. If the signal is large enough to be detected there is no problem, but what if it is not? You may be able to increase the sensitivity of the equipment, at a cost, and sometimes the strength of signal can be increased, but you will eventually be limited by the noise in the system. Noise has the effect of giving an output when no input is applied to the apparatus. It has the important characteristic of being random in time so that one cannot predict what the magnitude will be at any moment, only its average value over a long period.

The significant property of a system is not the size of signal, since this might be masked by a large amount of noise, nor the magnitude of the noise since this is irrelevant with a large signal, but the signal-to-noise ratio, S/N. Improvement in sensitivity must be directed at increasing this ratio. It is no use, for example, simply using greater amplification, since this will increase both signal and noise by the same amount, and the discrimination will not be improved.

One obvious method of reducing the effect of noise is to perform a difference measurement. First do a measurement with the signal, remembering that the noise is present also, so that you actually measure signal plus noise. Then remove the signal so that only noise is measured and subtract the two. There may be a problem if these two measurements are taken under different conditions or at largely different times since the noise level may not be quite the same in each case. An extended measurement should therefore be performed by alternating the two measurements at shorter intervals.

Another nice technique that can be applied if the signal has a regular time period is to sum the measurements made over many periods. The signal will steadily increase while the noise will tend towards zero, as at any instant it is just as likely to be positive as negative. Thus the S/N ratio steadily increases with the number of repetitions. This method can bring up a signal that at first is completely masked by noise.

Electrical noise is often caused by the random motion of electrons in components and this can be reduced by the simple expedient of cooling the offending part of the circuit. Again there is increased cost and inconvenience to be considered.

Units

Accurate measurement is only possible if your values can be compared with those performed in other laboratories so that checks can be made of the quality of your work, particularly in relation to systematic errors. To this end we need to define units of measurement, which represent the best that can be achieved at the time of their definition. The perfect unit is one that is very accurate and can be reproduced readily, so that every laboratory can have a copy. The five primary physical units described below fall short by different amounts from this ideal, and one has to make do with secondary standards of lower precision. It is the function of places such as the National Physical Laboratory (NPL) to maintain the best realization of each unit against which less demanding equipment can be compared.

(a) The *second* is the duration of 9 192 631 770 periods of the radiation produced in the transition between two energy levels in the caesium-133 atom. This is translated into a timescale synchronized to 0.1 μs, broadcast by radio and satellite transmissions.

(b) The *metre* is the distance travelled by light in vacuum in a time of 1/299 792 458 of a second. It is realized at the NPL using the wavelength of the 633 nm radiation from a stable helium–neon laser. Reproducibility is about 3 parts in 10^{11}.

(c) The *kilogram* is the mass of a platinum–iridium bar kept at the International Bureau of Weights and Measures in Paris. A copy is kept at the NPL, where masses may be compared to a precision of about a microgram on a precision balance.
(d) The *ampere* is the constant current that, if maintained in two straight parallel conductors of infinite length, of negligible circular cross section, and placed 1 m apart in vacuum, would produce between these conductors a force equal to 2×10^{-7} N m^{-1}. The standard is realized to about 0.08 μA using a current balance. The ohm is maintained to about 0.01 μΩ, and the volt to 0.01 μV.
(e) The *kelvin* is the fraction 1/273.16 of the thermodynamic temperature of the triple point of water. Water cells are used to reproduce the temperature of the triple point to 0.1 mK.

Examples of measurement of nuclear radiation

The three experiments described below illustrate the progression you might see in studying one topic in experimental physics, nuclear radiation. The first uses a Geiger tube to investigate whether the decay of a strontium-90 source follows the predicted Poisson distribution. The second shows that interfacing this equipment to a computer removes the tedium of collecting, displaying and analyzing the data. Finally, a scintillation counter is used to measure the energy and angular distributions of γ rays scattered from a metal rod in order to investigate Compton scattering.

Example 4.1 – the Geiger tube and Poisson distribution

Abstract

The correct operation of the Geiger tube was checked by observing that the graph of count rate against applied voltage showed the characteristic plateau. The operating voltage was set at 575 V, roughly in the middle of the plateau. With the source set to give a rate of a few counts per second, the number recorded over about 500 successive intervals of 1 s were recorded. The distribution in these values was plotted as a bar chart and shown to be a close fit to a Poisson distribution.

Aims

To calibrate a Geiger tube using a ^{90}Sr source and then use it to investigate the statistical law for radioactive decay, the Poisson distribution.

Introduction

A Geiger tube is probably the simplest device for detecting the presence of β particles arising from the decay of radioactive materials. It only counts the number of particles and gives no measurement of the energy which they deposit in the tube, unlike a scintillation counter. If the potential difference across the tube is high enough, the entry of a single β particle causes an avalanche of charge to be produced that leads to a measurable electrical pulse. This signal is of sufficient amplitude that a simple counter will suffice to record the number of pulses detected.

There is a range of voltages over which the counting rate is nearly constant. This plateau is a characteristic of the tube and enables a voltage to be set, about half way along, at which the count rate is virtually independent of applied voltage. Thus one does not need a highly stabilized power supply, with its consequent complication and expense.

The time at which a particular radioactive nucleus decays is not predictable; only the average number which decay in a specified interval. The random nature of the decay leads to the expectation that the distribution in the number of decays arising in a set time follows a Poisson distribution. This theory is tested in the current experiment.

Method

The source was a thin disc of ^{90}Sr of about 0.1 μCi strength, emitting β particles with energies up to 2.26 MeV. Simple precautions were taken to avoid skin contact with the source, and goggles were worn to avoid any damage to the eyes.

The Geiger tube was a Mullard type MX123, with cylindrical cathode enclosing a fine wire anode on its axis. A thin mica window at one end allows the entrance of β particles into the mixture of gases kept at a pressure below atmospheric; hence the concave shape of the window. Ionizing radiation entering the tube strips electrons from the gas atoms to leave positive ions behind. These charges are then accelerated towards the anode and cathode respectively, with the lighter electrons reaching speeds high enough to generate further ionization. The result is a cascade of charge sufficient to produce a measurable voltage pulse. The tube and source were mounted in a heavy brass castle to gain some shielding from background radiation, as illustrated in Figure 4.1. The circuit diagram is shown in Figure 4.2.

The tube was connected to a Mini-Assay unit to provide an adjustable high voltage source, measured with a digital voltmeter, and to count pulses for preset periods.

The correct operation of the Geiger tube was checked by measuring the count rate against tube voltage, at intervals of 50 V over the range 400–700 V. A few thousand counts at each point were enough to give a satisfactory indication of the plateau. This was achieved with a timing interval of 50 s.

To obtain a Poisson distribution the count rate was reduced to a few counts per second and the number recorded over a 1 s interval was repeated about 500 times.

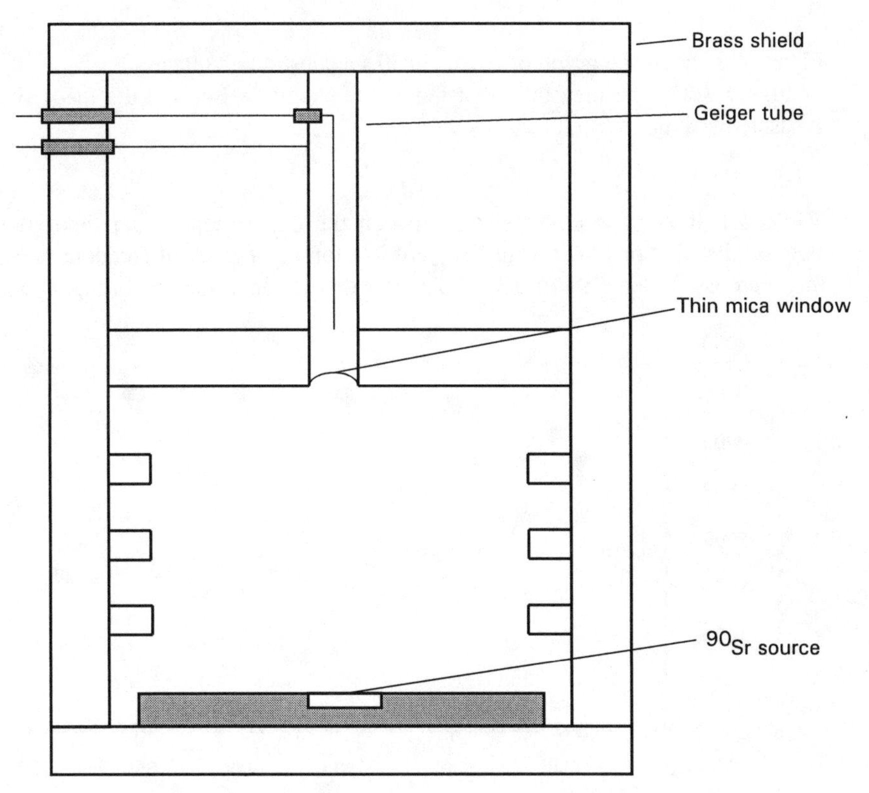

Figure 4.1 Geiger tube mounted in a brass castle.

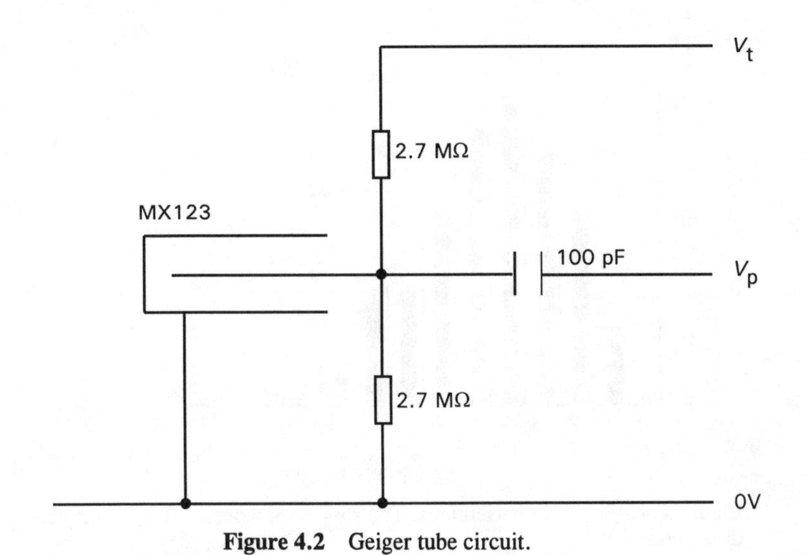

Figure 4.2 Geiger tube circuit.

Results

Figure 4.3 shows the graph of counts in 50 s against tube voltage.

Figure 4.4 is the measured distribution of counts in 1 s, and the theoretical Poisson distribution fitted to it.

Analysis

Table 4.1 shows the comparison between the experimental and theoretical Poisson distributions. The value of χ^2 of 7.4 for 13 degrees of freedom means that there is a probability of only about 10% that the agreement is brought about

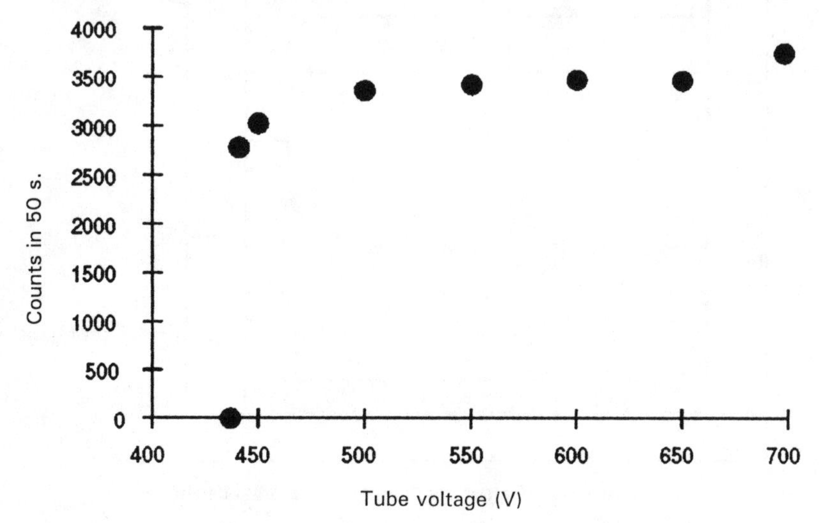

Figure 4.3 Geiger tube characteristics.

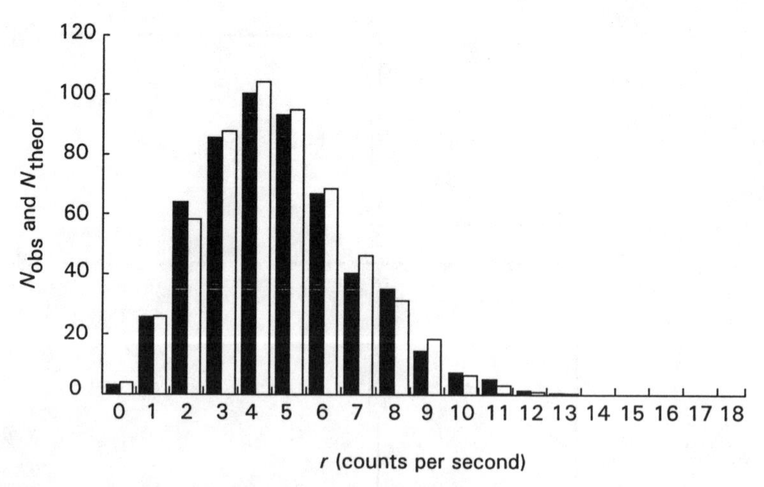

Figure 4.4 Poisson distributions. The solid bars represent the experimental distribution.

Table 4.1 χ^2 values for the Poisson distribution.

Pulses counted per second r	Number observed, N_{obs}	Poisson theory, N_{theor}	$r \times N_{obs}$	Contribution to χ^2
	5	5.4	0	0.030
1	25	25.0	25	0.000
2	68	58.2	136	1.650
3	87	88.7	261	0.033
4	98	102.3	392	0.181
5	93	93.6	465	0.004
6	69	71.8	414	0.109
7	41	47.3	287	0.839
8	31	27.2	248	0.531
9	11	13.6	99	0.497
10	8	6.5	80	0.346
11	5	2.7	55	1.959
12	2	1.1	24	0.736
13	1	0.5	13	0.500
14				–
15				–
16				–
17				–
18				–
Sums	544	543.9	2499	7.415
Mean			4.594	

by chance. For a Poisson distribution the mean value is equal to the square of the standard deviation, and the agreement between these values is reasonable (the mean = 4.6, σ^2 = 5.0).

Conclusion

The data obtained fits the model of a Poisson distribution to a probability of about 90%.

Example 4.2 – interfacing to a computer

This is a modification to the previous experiment to avoid the tedium of recording, displaying and analyzing the 500 or so measurements necessary for a Poisson distribution. As Charlie Chaplin demonstrated so effectively in his film *Modern times*, human beings are not happy performing simple repetitious activities. Computers on the other hand are precisely designed for such work, so why not ask them to do it?

It is not only that you can do what you did before with less tedium, and thus less chance of error. For a start, you can collect more than 500 points and thus achieve a smoother distribution. Secondly, you can perform experiments with a range of counting rates in order to see the transition from Poisson to Gaussian distributions as the mean count rate increases. These are typical advantages to be gained from interfacing your equipment to a computer. As in life you do not expect to get something for nothing, so there is more circuitry to build and depend on, and there is a possibility that the data will be less understood because you do not have to handle it as directly as before.

Basic method

First the pulse from the Geiger tube is given a "cleaner" shape so that it will be easier for the computer circuits to count reliably. This is shown schematically in Figure 4.5.

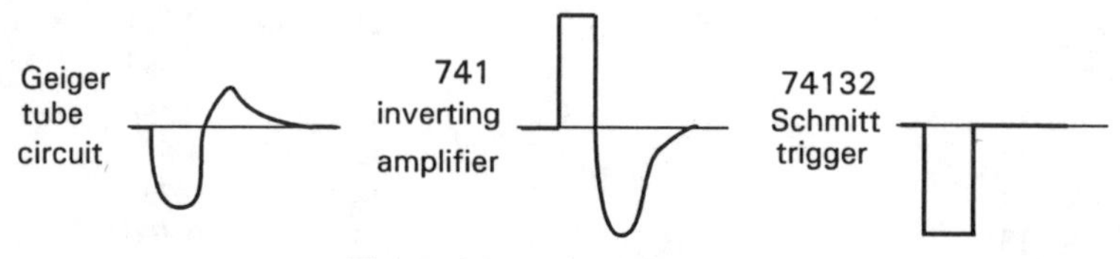

Figure 4.5 Pulse shaping.

The user port in a BBC microcomputer can be made to count negative pulses connected to pin PB6 by the following software arrangements. The pins on the user port have to be set to receive inputs rather than provide output signals. Then the timer T2 has to be set in counting mode. Next a register is made to store a number equal to the total counts to be registered, which is reduced by 1 every time a suitable pulse is received on pin PB6. When this number reaches 0 an interrupt flag is set to record the occurrence. Finally the keyboard buffer is flushed ready for the next measurement.

Example 4.3 – the Compton scattering of γ rays

This will be discussed in the form of a laboratory script presented to a student before performing the experiment. Since the knowledge of a group of students is by no means uniform, it is wise to give them ample time to read the script before performing the experiment. It is a matter of the writer's judgement how much background information needs to be included, remembering that the same script must cater both for students performing this as their first experiment, and those meeting it near the end of the laboratory course.

Aims

(a) To measure the change in energy of 662 keV photons scattered by electrons in a metal rod as a function of the scattering angle.
(b) To calculate a value for the rest energy of an electron from these measurements and to compare it with the accepted value.

Introduction

Is it a particle, is it a wave? No, it's an electron! The confusion over the nature of light and electrons is hardly clarified by regarding them sometimes as one and sometimes the other. If this seems unsatisfactory, remember that this is always the case when our understanding in science is incomplete, and we have to limp along with whatever understanding we have until someone explains the mystery for future generations.

The experimental basis for the wave nature of light began early in the nineteenth century when, for example, Young explained the interference effects created when coherent radiation passes through two adjacent narrow slits. It was more than 100 years later that the particle nature of electromagnetic radiation, in the form of X rays, was first demonstrated by Compton. A modern version of this experiment is your present assignment.

Compton found that radiation of a given wavelength passing through a metal foil is scattered in a matter inconsistent with classical theory, which predicts that the intensity observed at an angle θ will vary as $1 + \cos^2 \theta$ and be independent of the wavelength of the incident radiation. In fact, Compton found that the radiation scattered through a given angle consisted of two components: one the same wavelength as the incident radiation, the other shifted in wavelength by an amount depending on the angle of scatter.

Theory

Instead of thinking of the incoming radiation as a wave, Compton treated it as a beam of photons, each of energy $h\nu$. Individual photons in the beam were then scattered by electrons in the metal target, in a way analogous to billiard ball collisions. In such an elastic collision both momentum and energy are conserved, and a momentum is assigned to the photon of magnitude $p = h\nu/c$, where c is the speed of light. The target electron is assumed at rest, so its initial momentum is zero. However it does not have zero energy, as relativistic mechanics assumes a particle of rest mass m_0 has an energy m_0c^2 associated with it. Figure 4.6 shows the energies, momenta and angles involved in a typical collision, with the input data in bold print.

Conservation of energy:

$$E + m_0c^2 = E' + (m_0{}^2c^4 + p^2c^2)^{1/2} \tag{14}$$

Conservation of momentum along direction of incident beam:

$$E/c + 0 = E'/c\cos\phi + p\cos\theta \tag{15}$$

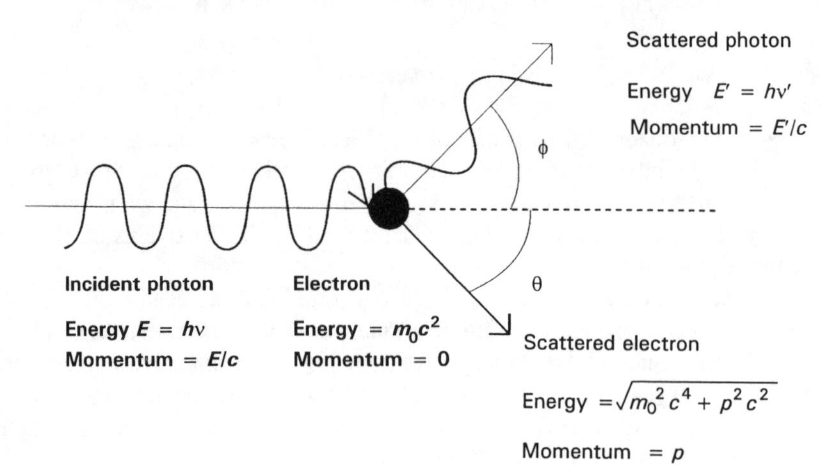

Figure 4.6 Scattering diagram.

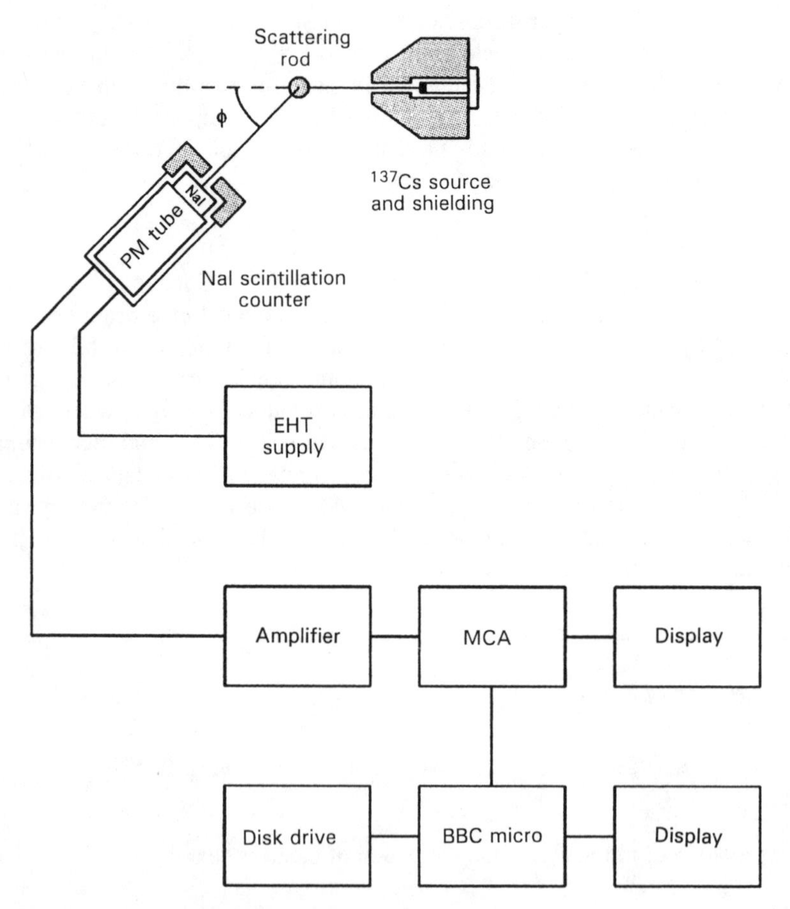

Figure 4.7 Arrangement of the apparatus (EHT, extra-high tension; MCA, multichannel pulse height analyzer; PM, photomultiplier).

Conservation of momentum perpendicular to direction of initial photon:

$$0 = E'/c \sin\phi - p \sin\theta \tag{16}$$

Since we detect the scattered photon, rather than the electron, we need an equation in ϕ and therefore eliminate θ from Equations 15 and 16:

$$p^2c^2 = E^2 - 2EE'\cos\phi + E'^2 \tag{17}$$

Again we eliminate from Equations 14 and 17 the variable that is not measured, p:

$$E' = E\left[1 + E/m_0c^2\,(1 - \cos\phi)\right]^{-1} \tag{18}$$

The experiment measures E, E' and ϕ, so that m_0c^2 can be calculated from the straight line graph obtained by plotting $1/E'$ against $\cos\phi$.

Apparatus

A layout diagram of the apparatus is shown in Figure 4.7.

Monochromatic 662 keV γ rays from the ^{137}Cs source are scattered by electrons in the metal rod and the energy of the γ rays is measured for a range of scattering angles, ϕ. The detector consists of a 50 mm diameter by 50 mm thick sodium iodide scintillator crystal optically coupled to a photomultiplier tube. Shielding the crystal from stray γ rays and the photomultiplier from stray light ensures that the signals sent to the amplifier are primarily the ones produced by the scattered γ rays. The detector can be rotated about the scattering rod through an angular range of 0–135°.

The light pulses (2 μs in duration) produced by the γ photons in the crystal are converted to a current pulse by the photomultiplier tube and then, by means of a fast amplifier, to voltage pulses that are then fed into a multichannel pulse height analyser (MCA). The MCA records the voltage of the input pulses, in one of 256 channels of equal width, according to their amplitude. The number of pulses directed to each channel can then be counted. The scintillation counter and MCA are described in more detail later.

Lead shielding around the source and detector, although heavily attenuating, cannot totally prevent some γ rays reaching the detector straight from the source. Thus there is always a background count that must be subtracted if the true signal due to scattering from the rod is to be obtained. To this end the data recorded by the MCA can be stored in the memory of a BBC microcomputer. Software is provided to derive the difference between two spectra, one with the rod in place and the other without.

Experimental procedure

After familiarizing yourself with the controls on the apparatus, you will need to set the EHT voltage applied to the photomultiplier so that the full range of the MCA is used to display the energy spectra of the γ rays and background. To do this, remove the rod, set the angle of scattering to 0°, and switch the MCA to record in 256 channels. Now carry out the following sequence, set out in the form of a loop in a computer program:

```
REPEAT   Set the EHT of the photomultiplier
         Count for about 30 s
         Read channel corresponding to the 662 keV peak
UNTIL    Peak channel is in the range 200–230
         Record the EHT and peak channel
```

The required value of the EHT is likely to be about 750 V.

Now that suitable conditions have been achieved, values of the scattered energy E' for a range of angles ϕ have to be obtained. The counting rate will be much smaller than during the setting up procedure ($\phi = 0$) so that longer times will be required to produce clear γ spectra. Again we can write the instructions in the mode of a loop in a computer program:

```
FOR
         φ = 15–135° STEP 15°
         Clear data from the MCA and computer memory
         Record spectrum with the rod present
         Store data in the BBC microcomputer
         Record the spectrum with the rod removed
         Display the difference between the two spectra
         Record the channel occupied by the scattered peak
NEXT φ
```

Calculations

Assuming the energy of the photons detected by the photomultiplier to be proportional to the amplitude of the voltage pulse and hence the channel number, calculate the scattered energies E' at each value of ϕ. Calibration is achieved by using the channel number corresponding to the 662 keV peak found during the setting up procedure.

Plot $1/E'$ against $\cos\phi$. A straight line graph should be produced, with the intercept on the $1/E'$ axis of $1/E + 1/m_0c^2$ and a slope of $-1/m_0c^2$. Obtain values for the rest energy of the electron, m_0c^2, from both the slope and intercept, and compare them with each other and with the accepted value.

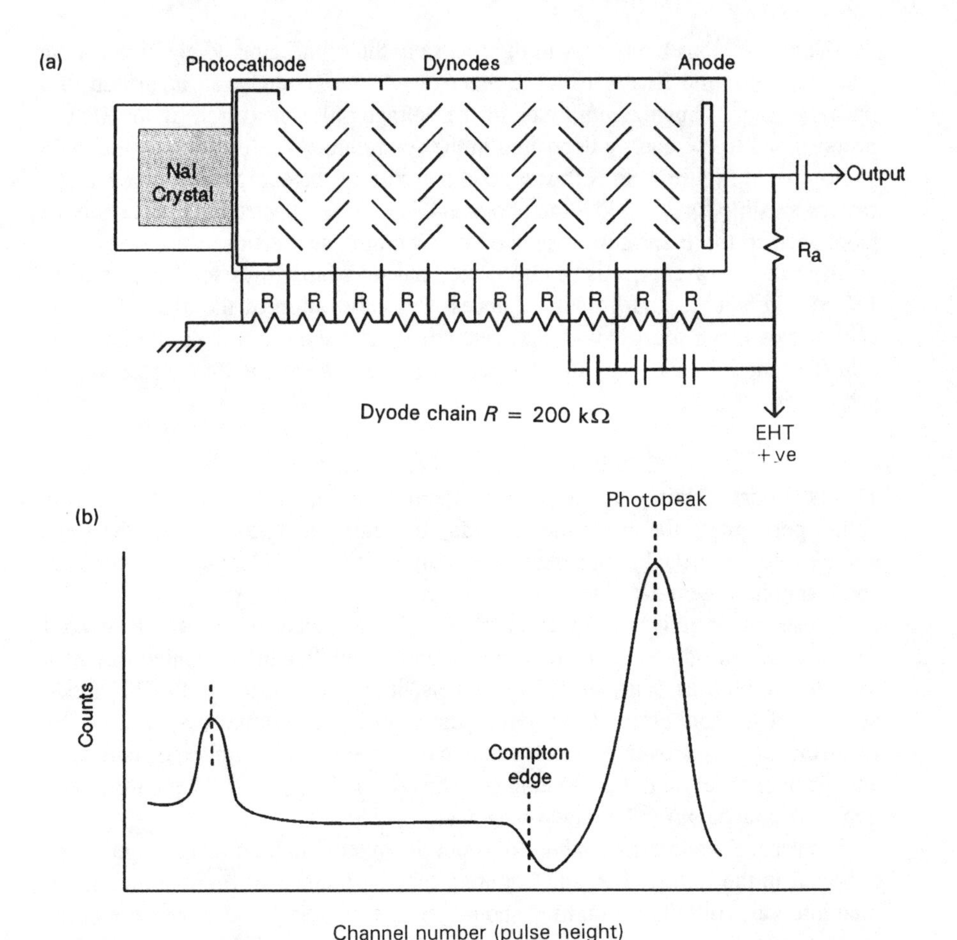

Figure 4.8 (a) A scintillation counter. (b) Pulse height spectrum for a caesium-137 source.

Appendix 1 – the scintillation counter

Figure 4.8a shows the arrangement of scintillator crystal, photomultiplier tube, and circuit connections that together make a scintillation counter for the detection of γ rays.

The scintillator is a crystal of sodium iodide doped with a small amount of thallium. It absorbs the incident radiation and converts most of the energy into pulses of visible light. At the photocathode this light is converted into a burst of electrons through the photoelectric effect.These electrons are then accelerated through a potential difference of about 100 V to the first dynode (which is just a fancy name for an electrode) where they have sufficient energy to produce secondary electrons in greater numbers. This multiplication occurs at each dynode until a pulse of perhaps 10^8 electrons arrives at the anode.

The conversions from γ ray to light pulse in the scintillator, to electrons at the photocathode, and finally to more electrons at the anode, are performed in a linear manner. Thus the amplitude of the voltage pulse measured by the MCA is proportional to the energy deposited in the scintillator. The position of the main photopeak in Figure 4.8b is thus a good measure of the energy of the 662 keV γ ray responsible for it. Below the photopeak is a broad spectrum of pulse heights from zero to the maximum produced by Compton scattering in the crystal. A small peak at low energy is often visible, arising from γ rays scattered through 180° in the source and holder. If necessary, calibration of the system can be achieved by using sources with different energies to show that the pulse height is a linear function of energy down to about 20 keV, a satisfactory range for the present experiment.

Multichannel pulse height analyzer

This is almost certainly a new piece of apparatus to you, so try out the controls before performing the experiment. It may be helpful to make a list of the steps you have to go through when using the equipment to avoid leaving one out later and losing hard won data.

As each pulse arrives from the amplifier, the MCA measures its amplitude and adds a count of one to the channel representing this height. Although the MCA can operate with as many as 1024 channels, it should be used in the 256 mode because of the moderate resolution of the scintillation counter. Spreading the spectrum of pulses over more channels would reduce the number of counts in any given channel, and the counting statistics would be poor (the standard deviation on N counts is $N^{1/2}$ for random events).

The vertical scale can be adjusted to suit the maximum number of counts to be expected in the fullest channel. Counting may be manual or preset to a chosen time interval. Automatic counting should be used so that the scattered and background spectra are obtained under identical conditions and comparison is thus fair.

To assist in identifying the channel in which a peak occurs, a cursor can be stepped along the channels with the channel number and its contents being displayed.

Computer program

This performs the transfer and storage of the scattered and background spectra from the MCA and displays each, and their difference, on a monitor. As with any respectable piece of software, instructions for its operation are displayed on the monitor as required. If you are interested, you could make a copy of the program on a printer, to follow its instructions.

CHAPTER 5

Analysis of data

Statistical thinking will one day be as necessary for efficient citizenship as the ability to read and write.

H. G. Wells

Round numbers are always false.

Samuel Johnson

To understand God's thoughts we must study statistics, for these are the measure of his purpose.

Florence Nightingale

Single experimental variable, repeated *N* times

Introduction

Although practical considerations place an upper limit on the number of measurements we are likely to make, in principle there is no limit to the number. This situation is described by saying that the size of the *population* is infinite. The actual number of measurements made is called the *sample*, and is clearly of finite size, N. What we are trying to do in experiments is to make a good estimate of the properties of the population, while only making a finite number of readings. At first sight this may appear an impossible dream since you do not need to be much of a mathematician to realize that 10 or 20, say, is a long way short of infinity. Fortunately, it turns out that we can indeed get a reasonable understanding of the population from a sample of such a size. Clearly it will not be perfect knowledge, but scientists can do wonders with incomplete information!

What value should we choose for N?

$N = 0$ This is clearly unsatisfactory since there is not even an attempt at measurement, though the absence of knowledge does not prevent some people expressing an opinion of course.

$N = 1$ Better than nothing, but it does not even provide a check on mistakes, let alone give us a measure of the variability in the measurement. However, there are circumstances when we are happy to accept these limitations: no one wants a repeat of the accident at Chernobyl to investigate the effects of nuclear radiation on the world's environment.

$N = 2$ A further improvement, since it allows a check on any mistake in the first measurement, unless you make the same mistake twice! Also, it is the smallest number to provide some idea of the variability in the measurements. Still not satisfactory, though, since the measure of variability is too crude.

$N = 10–20$ Now we are talking! This is a reasonable compromise if the measurements have to be performed manually. We have to balance the desire for an even higher number against the time needed to achieve it. Common sense suggests that the longer an experiment proceeds the greater is the chance of systematic errors or mistakes intruding.

This number is enough to obtain a reasonable estimate of the variability of the population, from which it is a typical sample, we hope. However it is still not large enough to produce a smooth bar chart or histogram (see Fig. 1.1), so we cannot test whether the distribution is Gaussian, as is often assumed.

$N >> 20$ Such large numbers can be achieved if the apparatus is automated by linking it to a computer; an example is discussed in Chapter 4. Now a histogram can be plotted, with some confidence that the shape of the distribution can be investigated. It is wise to plot a time series of the data (Fig. 1.2), particularly if the experiment takes a long time, to see whether there is evidence of any drift with time.

Best estimate of the true value

Given N measurements of the experimental variable, we want to make a summary of the data that gives the best information about the population. The first element in this summary is the true value, which is approximated by the arithmetic mean. This is the simplest example of the application of the *principle of least squares*, which we will use again later. This simply says that the best value is the one that minimizes the sum of the squares of the deviations of each value from the best. If we write the sum as S, the best value as B, and consider N values of the experimental variable x, the analysis is as follows:

$$S = \sum_{1}^{N} (B - x)^2 \qquad \text{is a minimum}$$

$$\therefore \frac{dS}{dB} = 0$$

i.e.

$$2\sum_{1}^{N} (B - x) = 0$$

$$\therefore NB = \sum_{1}^{N} x$$

and

$$B = \frac{1}{N}\sum_{1}^{N} x = \bar{x} \qquad \text{the mean.} \qquad (19)$$

That is, the best estimate of the true value of the population is the arithmetic mean of our sample, always assuming there is no significant systematic error in the measurements.

A measure of variability in the measurements

I am sure you could invent your own measure of variability, since there are many possibilities, but scientists have to agree a common definition so that we all speak the same language. Also, we might as well use the definition of the sum, S, used above, as it is consistent with the principle of least squares. This makes the definition of the *standard deviation* of our sample, σ_s, as shown below. Since we are in fact trying to discover the properties of the population from which our sample was taken, we also need to define our best estimate of the standard deviation of the population, σ_p. The difference arises from the fact that one piece of information, the mean, has been taken from the N measurements, leaving $N - 1$ pieces to calculate the standard deviation:

$$\sigma_s \equiv \left[\frac{S}{N}\right]^{1/2} \qquad \text{(called } \sigma_n \text{ on calculators)}$$

where

$$S = \sum_{1}^{N} (x - \bar{x})^2$$

and

$$\sigma_p \approx \left[\frac{S}{N - 1}\right]^{1/2} \qquad \text{(called } \sigma_{n-1}\text{)} \qquad (20)$$

How close is the mean to the true value?

If we keep increasing N, after a while the values of the mean and standard deviation do not vary very much. But there must be virtue in increasing N further, since each measurement gives us an extra piece of information. The improvement arises in the third of the numbers used to summarize our data, the *standard error of the mean*, α, which is defined below:

$$\alpha = \frac{\sigma}{\sqrt{N}} \tag{21}$$

In the absence of systematic error in the measurements, this estimates how close the arithmetic mean is to the true value. If you know what the true value is then you can observe α getting steadily smaller as N increases (see Example 2 below). If you do *not* know what true value to expect, then there is always the possibility that you have a systematic error, which makes your mean either low or high. A good way of checking for this possibility is to compare your measurements with those of someone else; it is unlikely that you will both have made the same systematic error.

Example 5.1 – mean, standard deviation, standard error

In Table 5.1 you will see data laid out to enable you to calculate the mean, standard deviation, and standard error of the mean of a general experimental variable x.

Table 5.1 Mean, standard deviation and standard error of the mean.

Reading	x	(x – mean)	(x – mean)^2
1	2.00	0.003	0.000 009
2	2.03	0.033	0.001 089
3	1.98	–0.017	0.000 289
4	2.01	0.013	0.000 169
5	1.95	–0.047	0.002 209
6	2.02	0.023	0.000 529
7	2.01	0.013	0.000 169
8	1.97	–0.027	0.000 729
9	2.01	0.013	0.000 169
10	1.99	–0.007	0.000 049
Sum	19.97	0.000	0.005 410
Mean	1.997		
Standard deviation			0.025
Standard error			0.008

In column 1 I have numbered each reading so that there is a record of the order in which they were taken.

Although it is not specifically stated in column 2, by writing the numbers to three significant figures (two decimal places if you prefer) it is *implied* that the accuracy of measurement is a few parts in the third figure. It is far better to clearly state your estimate of uncertainty so no doubt can arise. I have only omitted it for clarity.

You may be surprised that the mean value is quoted to four figures when the individual values are to three only. This is because, by taking ten readings, you have gained greater accuracy for the mean than is present in any one value. This is the pay-off for hard work, so do not shun it!

Column 3 is not necessary for the progress of the calculations, since you could easily go straight from column 2 to 4 without loss of information. It is put there to show that the sum of that column is zero. This arises directly from the definition of the mean, of course, but it is often a good idea to have some way of checking that computations are progressing correctly before reaching the end. If you had worked out the wrong value for the mean, the sum in column 3 would not be zero and you could correct the mistake without carrying it into later calculations. Such an idea is easy to include in a computer program, of course. You will meet it again when we apply the principle of least squares to fitting the best straight line to a graph. Because of rounding effects the sum may not be precisely zero, as in this case, but must be so within rounding errors.

Column 4 enables us to calculate the standard deviation, which we quote to three decimal places to be consistent with the mean. Note that the magnitude of the standard deviation is the same order as the fluctuations in the measured values. Such a common-sense check is always worth making, since we can all make mistakes in calculation or typing numbers into a keyboard. Note that we have had to use six decimal places, to avoid any rounding problems. If you do not see the point, try the calculation with only three decimal places in column 4.

We can now calculate the standard error of the mean. Quoting it to one significant figure leaves us with three decimal places for the mean and its error, which is consistent. You may sometimes see the error quoted to two significant figures, but never more. The mean would then be written to four decimal places, to be consistent, i.e. 1.9970 ± 0.0078.

The standard deviation is a measure of the variation of the experimental measurements about the mean, but does not represent the outer bounds of values. Reference to Table 5.1 shows that observations 2, 5 and 8 lie outside the range ± 1 standard deviation about the mean. This is quite normal, and will be discussed in more detail later.

Quite often students confuse the ideas of significant figures and number of decimal places. I have used each in the description above so that you become familiar with each.

Example 5.2 – mean, standard deviation, standard error: time series

This is for the gamblers among you! Traditionalists will use the time-honoured six sided die, while the modernists will share my preference for using random numbers generated by computer. I have included an outline of my program "DICE" in an appendix in case you wish to use or develop it further.

I wanted an example for which the results were known, so we would not have to worry about systematic errors biasing the results, as they often do in measurements. The procedure is to throw the die a large number of times (100 in my example), and after each throw calculate the new values of the mean, standard deviation, and standard error of the mean. Because these values are changing I have called them the "running" values. If the die is fair, or the random number generator valid, each number (1–6) is equally likely to occur at each throw. The true value that would be obtained with an infinite number of throws is then the arithmetic mean of the six numbers, 3.5. With our sample of 100 throws we will be a little away from this value for the infinite population, but the expected value is clear. Similarly the standard deviation of the infinite population will be 1.71 to three significant figures. Since we are going to use 100 values only, the standard error of the mean will be 1.71/10 = 0.17 to two significant figures. To summarize, the true values we are expecting are

Mean	3.50
SD	1.71
SEM	0.17

In Table 5.2 you see the results for each random number (1–6) up to 20, and every 10 afterwards. When N is small, all three calculated values vary appreciably, but settle down quite soon. Even after only ten random numbers the values are only slightly high.

Naturally, the deviations are even smaller for $N = 100$ (0.04, 0.06 and 0.01, respectively) but the improvement in standard error of the mean is not as rapid as you might have thought. This arises from the factor $1/N^{1/2}$ in the calculation. Changes in N are more effective when N is small, and less so as N rises. This is an example of the law of diminishing returns: the more measurements you have taken, the more you *have* to take to gain a particular improvement.

As N increases note that both the mean and standard deviation tend to constant values, the latter because we are making roughly equal contributions to both the numerator and denominator in the defining equation. The standard error, however, steadily decreases, as shown in Figure 5.1, the time series for the same data.

Note that the range represented by ±1 standard deviation about the mean (1.63–5.37) causes all the ones and all the sixes to lie outside it. Having one-third of the data outside this range is perfectly normal.

Table 5.2 Results of 100 generations of random numbers (1–6).

	——Running values——		
N	Mean	Standard deviation	Standard error
1	5.00	–	–
2	3.00	2.83	2.00
3	4.00	2.65	1.53
4	3.25	2.63	1.31
5	3.20	2.28	1.02
6	3.33	2.07	0.84
7	3.29	1.89	0.71
8	3.50	1.85	0.65
9	3.44	1.74	0.58
10	3.70	1.83	0.58
11	3.73	1.74	0.52
12	3.50	1.83	0.53
13	3.46	1.76	0.49
14	3.43	1.70	0.45
15	3.27	1.75	0.45
16	3.44	1.82	0.46
17	3.53	1.81	0.44
18	3.44	1.79	0.42
19	3.58	1.84	0.42
20	3.50	1.82	0.41
30	3.70	1.80	0.33
40	3.47	1.75	0.28
50	3.56	1.73	0.24
60	3.42	1.77	0.23
70	3.60	1.81	0.22
80	3.44	1.82	0.20
90	3.52	1.79	0.19
100	3.54	1.77	0.18

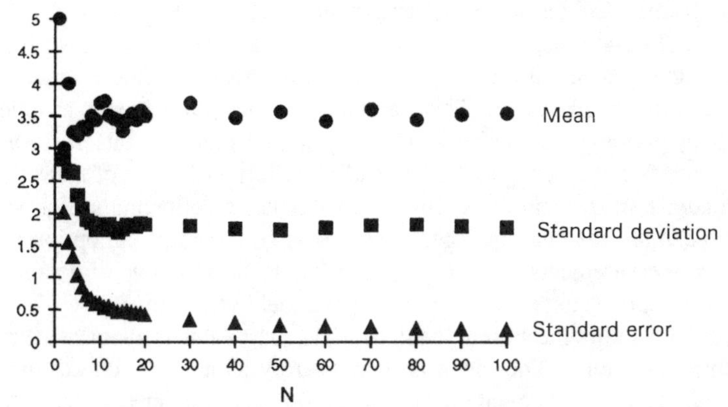

Figure 5.1 Time series for random numbers 1–6.

It is wise to develop a sceptical approach to apparatus, and not simply assume that it is functioning as you think it ought to. In this case, is your die fair, or the random number generator in my computer valid? We have assumed that each of our six numbers is equally likely to occur. A simple test is to generate many more than 100 numbers (sorry about you die throwers) and see whether they occur with equal probabilities. I wrote a simple program to count the number of times each of the six random numbers occurred in a total of 6000 events, so each number should show about 1000 times. The results are shown in Table 5.3.

Table 5.3 Distribution of 6000 random numbers.

Random number	Occurrences		
1	973	1002	941
2	916	1033	1043
3	1085	1036	959
4	1001	968	1031
5	1015	983	1011
6	1010	978	1015

On the first run there seem to be too few twos and too many threes, justifying a second trial. This still seemed to indicate too many threes, so a third trial was made. This showed clearly that the earlier numbers had been random fluctuations and not bias in the random number generator.

Example 5.3 – reaction time measurements

This uses a computer program, called "REACTIM", which I have written in BBC BASIC and 6502 assembly language, to measure reaction times to a change in visual stimulus (outline listing 4.2 in Appendix 4). The timing has to be in milliseconds to obtain sufficient discrimination, which requires the use of assembly language rather than the slower timer available in BBC BASIC.

You could make the visual stimulus more elaborate if you wish (colours, flashing lights, etc.), but I have simply used the transition, after a random time, from a full to a blank screen. This starts a timer that you then stop as quickly as possible by pressing the space bar. The program contains two sets of stored data, which are referred to as "good" and "bad": the former is a set of measurements taken under normal conditions while the latter has a deliberately high value so that you can see the consequences. There is also, naturally, an option to make your own measurements. The display is in the form of a table of data and time series, with options to print a copy on paper should you wish.

Figure 5.2 shows the stored "bad" data. The last value is about twice as large as the previous nine. The mean is increased by about 10% due to the rogue value, but the standard deviation by about 1000%. The latter is so large because not only does the rogue point give a large contribution to the sum of squares of

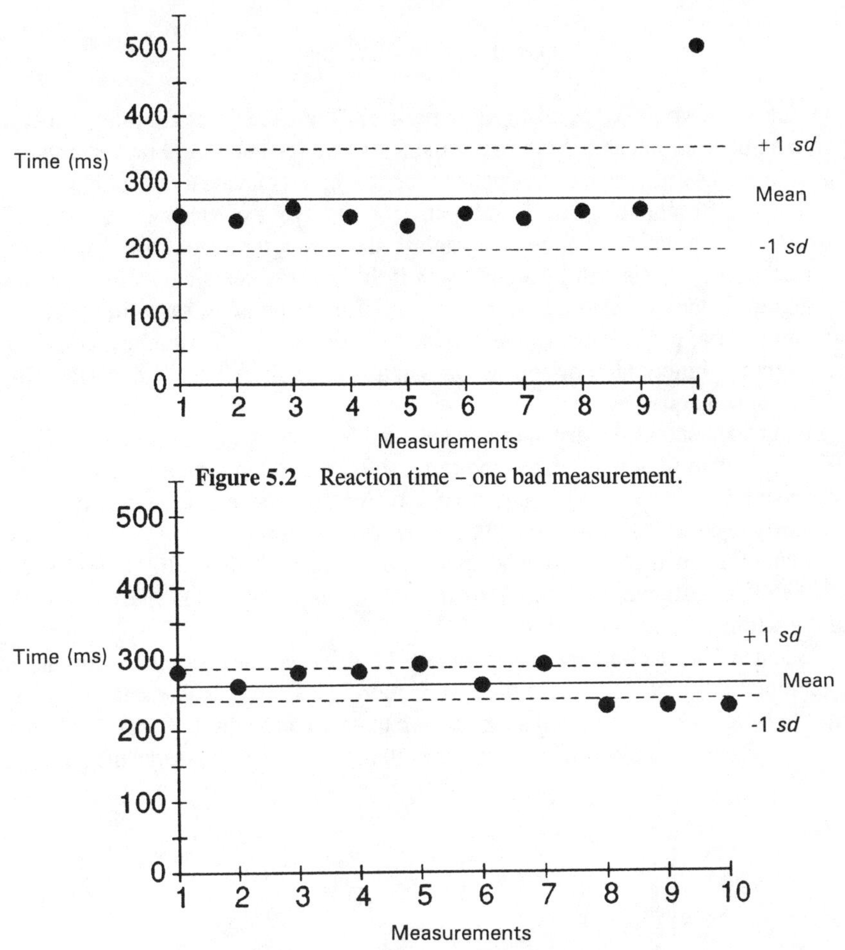

Figure 5.2 Reaction time – one bad measurement.

Figure 5.3 Reaction time – ten good measurements.

the deviations from the mean, but the displacement of the mean (see the time series) causes the other nine points to give contributions larger than usual. The time series also "looks wrong", since for good data we would expect about half the points to be below the mean and half above, instead of the 9/1 division here. Finally, it is normal to have about one-third of the points outside 1 standard deviation from the mean, rather than the single value here. Be on the lookout for such effects in your own measurements, though they are not likely to be so pronounced as here.

Many of the comments above do not apply to the data shown in Figure 5.3, which are my measurements. Approximately one-third of the points are outside the 1 standard deviation range, and half the points are above the mean, so the data seem acceptable. Why not make your own measurements, but if they turn out to be low, remember that quick reaction is a necessary, but not sufficient, attribute for a Formula 1 racing driver!

Distribution of values

On occasions you may perform an experiment for which it is easy to collect a large number of values of your experimental variable. This could be because the measurements are very easy to perform, or you have connected the equipment to a computer for collection of data. Apart from the types of analysis already discussed, we are now able to plot a graph to show how the values are distributed about the mean. This will be a bar graph if the bins contain single values, and a histogram if they contain a range of values. The shape gives extra information about the data, particularly if it is a close fit to one of the distributions common in science - binomial, Poisson or Gaussian - since these can be described by mathematical equations.

Figure 5.4 shows the distributions obtained by plotting the sums of 1, 2, 4, 8, 16 or 32 random numbers from the range 1–6. When the random numbers are taken singly, none is more likely to arise than any other, and a rectangular distribution occurs. The vertical scale, N, is just the number of occurrences per bin, out of the total of 10 000. The area under the graph is thus 10 000, as it is for all the others, allowing for the normal fluctuations expected in "measurement" and my limited skill in drawing.

When pairs of numbers are summed the distribution becomes triangular. Combined in greater numbers the distribution tends towards Gaussian, having the characteristic inverted bell shape shown towards the right in Figure 5.4. This is an example of the *central limit theorem*, which says that, whatever the starting

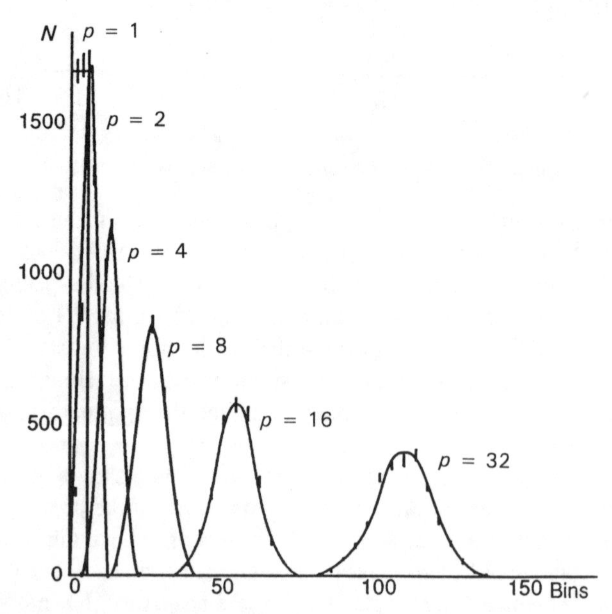

Figure 5.4 Sum of 10 000 random numbers taken p at a time.

distribution (rectangular in this case), the distribution formed by summing enough values tends to become Gaussian. Then we can use the Gaussian equation to specify the probability with which particular values are likely to arise. This topic will be discussed in more detail later.

Weighting of data - introduction

Quite often we have data of the same physical variable taken under different conditions (experimenter, apparatus, method, etc.) which we want to pool to get a larger set of data. How should we combine the separate measurements to give an overall mean, standard deviation, and standard error of the mean? The principles seem obvious: the more accurate a measurement, the greater contribution it should have in the calculations; and the weighting factor should be based on the principle of least squares. The steps leading to the final calculation are discussed below.

Step 1

Imagine we have 26 students, labelled A to Z, each of whom has made a single measurement of the periodic time of an identical pendulum. We might think that one student has probably made a better measurement than another because he or she is usually more careful, or has a better "feel" for measurement. This may tempt us to give greater weight to the result from the "better" student, when we combine them all, but this is too qualitative an approach to be acceptable. Without any numerical evidence to the contrary we must treat all 26 measurements equally. Of course, it would be sensible to look at the values to see if there is any obvious mistake in any of them.

So we calculate the arithmetic mean, standard deviation, and standard error of the mean in the usual way, treating all 26 values equally. We would then quote the class result as the mean ± the standard error.

Step 2

Of course the students should each have made a number of measurements of the time period, for the reasons discussed earlier in the chapter. Then each would have a mean and standard error of their own. There is no need for all of them to make the same number of repetitions since the number is reflected in their standard error; the ones who make more measurements get a smaller standard error, other things being equal. Clearly measurements with small error are better than those with larger errors, either because more measurements have been made or the data have a smaller spread. These should be given a greater weighting in the summary of the class data.

Step 3

The weighting factor is given by

$$W = 1/\alpha^2 \tag{22}$$

(You would expect to have an inverse function of α so that the better data have greater weighting, and the square is in line with the ubiquitous method of least squares - so there is no surprise here.)

The class mean is now called a *weighted mean* and is calculated by

$$\bar{T}_{\mathrm{w}} = \sum WT / \sum W \tag{23}$$

while the error on the weighted mean, $\bar{\alpha}_{\mathrm{w}}$, is given by

$$1 / \bar{\alpha}_{\mathrm{w}}{}^2 = \sum 1 / \alpha^2 \tag{24}$$

Example 5.4 - weighting: identical measurements (a class of identical robots)

This is clearly an artificial situation, but useful none the less in showing a limiting case. We assume that all 26 students have obtained identical values for the time period of 1.00 ± 0.01 s. Substituting these numbers into the equations above yields the class weighted mean to be 1.00 ± 0.002 s. Obviously the mean time period is that found by each student, and increasing the data set from 1 to 26 elements simply causes the error to be reduced by the square root of 26.

Example 5.5 - weighting: different measurements (a class with Michael Faraday and a bunch of idiots)

Again this is an extreme case, but it allows us to see what happens to the weighted mean when one piece of data is so much better than the rest. For simplicity we will assume that the 25 idiots all achieve the same result of 2.0 ± 0.2 s, while Faraday obtains 1.00 ± 0.01 s.

Before calculating the weighted class result why not give your scientific intuition free reign to consider whether you expect the result will be nearer one of the values quoted above or the other. On the one hand it is always better to have 25 pieces of information than 1, so you may favour the former value. This would be true if all the information were equally reliable, but Faraday has quoted an error that is 20 times smaller. This does not make his measurement just 20 times better, but 400 times, as we deal with squares of the errors. So we expect the single "good" result to have much greater weight than any one of the poorer ones. We have to do the calculation to see how much it is modified by the larger number of poorer readings.

The weighted class result is 1.059 ± 0.010 s.

This is much closer to the single good result than the 25 poorer ones. The lesson is clear: *quality first, quantity second.*

Two experimental variables

Introduction

Many experiments in physics involve two experimental variables that can be used to draw a straight line graph. Sometimes they can be used directly, as with distance and time for a body moving at constant speed; on other occasions you have to perform a little algebra first, as with the length and time period of a pendulum.

Are they correlated?

Look at Figure 5.5 overleaf. The diagrams show two extreme situations, followed by one that is more realistic.

In Figure 5.5a there is no evident relationship between x and y. They are said to be not correlated, so that being given a value for x will not allow you to predict the corresponding value for y, and vice versa. This is reasonable when I tell you that both were chosen as random numbers in the range 1–10. The result is sometimes called a scatter diagram because it is what you would get by letting objects fall onto squared paper. For a finite number of points you do not get the theoretical value for the correlation coefficient, 0 (see Appendix 2), but it is not far away. Drawing a straight line through these points would be rather silly, but if you did, the slope and intercept would not be far from their expected values of 0 and 5.5, allowing for sampling errors.

The graph in Figure 5.5b is formed from a mathematical relation $y = x + 1$, so it is a straight line, and we can make perfect predictions for one variable given a value for the other. The correlation coefficient is thus exactly 1, with the slope and intercept both 1, as expected. If the graph had a negative slope, such as $y = -x + 1$, the correlation coefficient would have been -1. These situations rarely arise in experiments because of random fluctuations in measurements, but they show the limiting conditions.

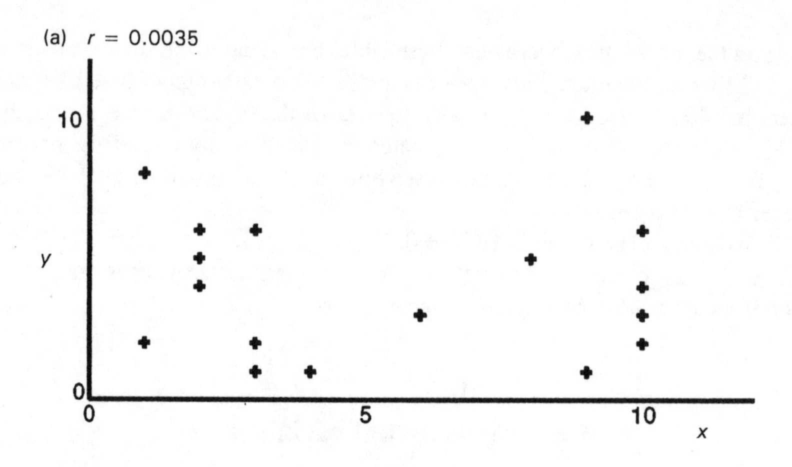

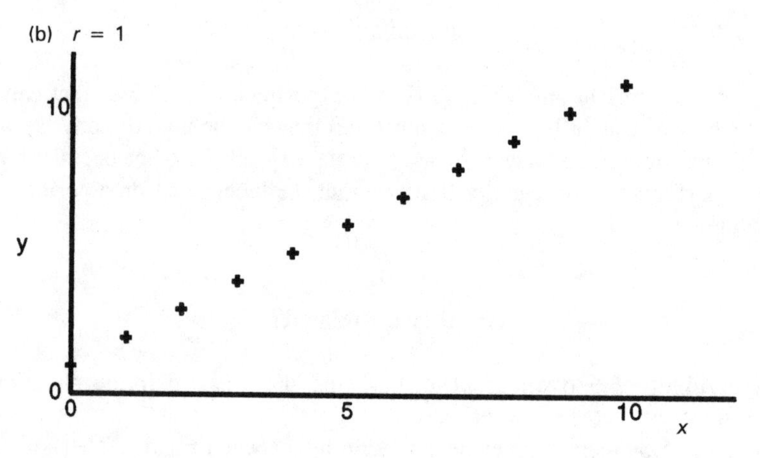

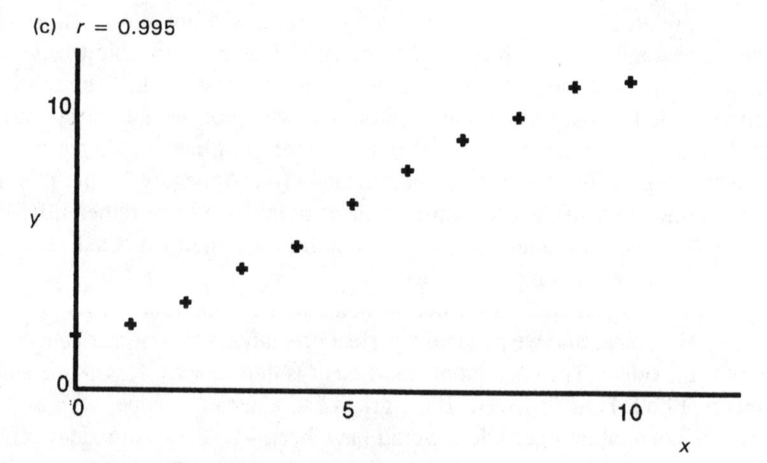

Figure 5.5 Various correlation coefficients r.

The graph in Figure 5.5c shows a situation that is more likely to arise in practice, where we have an expected linear function with a random fluctuation superimposed. The equation used to generate the data was $y = x + 1 + \text{RND}(1)$, where the last term is a random number between 0 and 1. The correlation coefficient is a little less than 1, and the slope and intercept close to their expected values of 1, allowing for the random errors.

So, the magnitude of the correlation coefficient indicates how accurately you are able to *predict* the y value corresponding to any new value of x. With $r = +1$ or -1 your forecast is perfect; with $r = 0$ it is pure guesswork; in between you are able to predict with an element of uncertainty depending on the value of r. I am tempted to suggest these three regions are those applicable to religion, politics, and science, respectively, but will the editor allow such flights of the imagination?

The trouble with using the correlation coefficient as a measure of how close data are to a straight line is that r does not vary much from 1, unless the data are so poor that you are unlikely to think of plotting a straight line, anyway. Physicists prefer to use the errors on slope and intercept as measures of how far data depart from a perfect line. Also, the slope and intercept are usually related, by theory, to significant quantities in the experiment.

You can investigate the relation between the linear correlation coefficient, r, and the slopes, intercepts and their errors in the following way. Obtain a series of approximately straight lines by using the equation $y = 1 + x + [n \times \text{RND}(1)]$ with n taking the values 1, 2, 3, etc. In each case perform a least squares fit to the data, using the WLSF program discussed in Appendix 4. As n increases the random component of the equation increases, with effects on r, m, c, σ_m and σ_c, which can be investigated quantitatively.

Fitting data to a straight line

Introduction

If there are pairs of experimental variables (x, y) which there is reason to believe are linearly related, we want a computer program that will calculate the slope and intercept of the "best line" (see program summary WLSF in Appendix 4). This is the line formed by applying the principle of least squares to the problem. The principle is a little more complicated to apply than when we used it to arrive at the arithmetic mean as the best estimate of the true value, but employs exactly the same ideas. As a bonus we also obtain estimates of the errors on slope and intercept, as well as the correlation coefficient if required.

Before performing the algebra it is worth looking at the steps that have to be gone through:

Step 1. Plot the experimental points on a sheet of graph paper so that you can judge whether a straight line is relevant; the computer cannot make this decision for you. It is unwise to assume a straight line to be appropriate simply because theory tells you to expect one; the theory may be wrong, one or more data points poorly measured, etc.

If each of your points is the mean of many readings, do not forget to plot error bars of length $\pm\sigma$ about the mean. These will help you to decide whether a point off the line is due to acceptable random errors or whether there are systematic errors that should be reduced before continuing the analysis. And remember that one-third of your points may be more than one standard deviation from the mean.

Step 2. The computer program is divided into four cases, depending on your knowledge of the errors on the variables x and y. You must select the one suitable for your data.

Step 3. Type your data into the keyboard, using the format specified in the program. You will be given the opportunity to check the data, amend it, or remove items before initiating the calculation of the best line.

Step 4. Read the values of the slope, intercept, and the errors on each, calculated by the program. Round the errors to two significant figures, and then the slope and intercept to the same number of decimal places.

Step 5. Using these values of slope and intercept calculate two well separated points on the line $y = mx + c$ and draw the straight line connecting them. This should look a reasonable fit to your points. If it looks obviously wrong then a mistake has been made and must be identified and corrected before proceeding.

Step 6. The program also calculates a "centre of gravity" point (x, y) through which the line must pass if it is correct. Plot this point, using a different symbol from your experimental points, as a further check on the correctness of the calculations.

Step 7. Decide whether any point is so far from the line, relative to its error bar, if it has one, that it may have arisen from a mistake or systematic error in the measurement. Check the measurement if possible, and either remove the point from the calculation or change its value. Return to step 3 and repeat the sequence.

The four cases

1. Each point on the graph arises from single measurements of x and y. There is thus no knowledge of the uncertainty in either experimental variable (σ_x and σ_y are unknown). This situation is quite common, though the arguments in favour of repeating readings are as valid here as they were for a single variable. (This is the case used on those calculators programmed with a "linear regression" function.)

2. At each value of x, repeated readings are taken of y so that we are able to calculate σ_y at each point. For case 2 we assume that all values of σ_y are equal. We still assume no knowledge of σ_x.
3. This is a slightly modified version, in which the values of σ_y may vary from point to point.
4. This most general case is that in which we have values for both σ_y and σ_x which are comparable in size. It could be used for all four types of calculation by inserting standard deviations of zero, as appropriate, though the calculations are more difficult and it seems a waste of time typing zeros into the keyboard. Use it when needed but not as a substitute for the simpler, earlier cases.

Cases 1–3 assume that any deviation of an experimental point from the best line is due to error in the y variable only, the x measurement being accurate. You might think this is a bit rich in case 1, for which we have no knowledge of errors in either x or y. Well, not even scientists are perfect! There is one important consequence of this assumption: *you must plot the variable with the greater error on the* y *axis*; the algebra on which the computer program is based requires this condition. This is contrary to any advice you may have received that the independent variable is to be placed on the x axis, with the dependent variable on the y axis.

Analysis of case 1

We will look at the analysis of this case in some detail, and deal in broader terms with the changes required for the other three cases.

Figure 5.6a shows the best straight line, $y = mx + c$, with a representative point (x_i, y_i) at a distance d_i from the line. The displacement of the point from the line is assumed to arise from errors in y, as explained earlier.

The displacement d_i is the difference between the y value at the point, and that on the line directly below it:

$$d_i = y_i - (mx_i + c) \tag{25}$$

This is positive for the point shown, but would be negative for a point below the line, as is equally likely if the line is a valid fit to the experimental data.

The principle of least squares requires us to calculate the values of m and c which minimize $S(1)$ (the sum for case 1) the sum of squares of d_i, i.e.

$$\begin{aligned} S(1) &= \sum (d_i)^2 \\ &= \sum (y_i - mx_i - c)^2 \end{aligned} \tag{26}$$

where the summation occurs over all the experimental points.

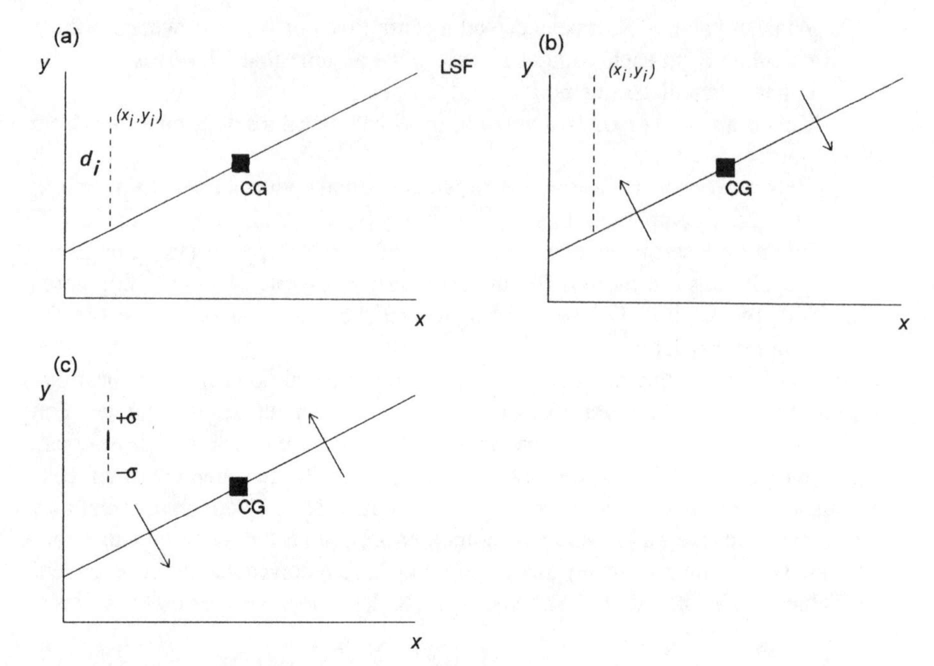

Figure 5.6 Line of least squares fit (LSF), centre of gravity (CG) and typical experimental point (x_i, y_i). (a) Typical point at distance d_i from the line. (b) If y_i increases, the best line rotates (clockwise) to get closer to the point. (c) If the error bar increases, the weighting of the point decreases, and the line rotates (anticlockwise) to get further away. The changes in the second and third graphs also effect the position of the CG, but this is a minor effect.

To find the minimum we must differentiate first with respect to m, keeping c constant, and then with respect to c, keeping m constant, and setting each differential equal to zero:

$$\begin{aligned} 0 &= \left(\partial S(1) / \partial m\right)_c \\ &= \sum 2(y_i - mx_i - c)(-x_i) \\ &= \sum x_i y_i - m\sum x_i{}^2 - c\sum x_i \end{aligned} \tag{27}$$

and

$$\begin{aligned} 0 &= \left(\partial S(1) / \partial c\right)_m \\ &= \sum 2(y_i - mx_i - c)(-1) \\ &= \sum y_i - m\sum x_i - Nc \end{aligned} \tag{28}$$

where N is the number of points on the graph. Dividing through the last equation by N shows that the centre of gravity point $(\bar{x}, \bar{y})$ must be on the line, i.e.

$$0 = \bar{y} - m\bar{x} - c \tag{29}$$

Solving Equations 27 and 28 for m and c gives

$$m = \frac{N\sum xy - \sum x \sum y}{N\sum x^2 - (\sum x)^2} \tag{30}$$

$$c = \frac{\sum y \sum x^2 - \sum x \sum xy}{N\sum x^2 - (\sum x)^2} \tag{31}$$

Uncertainty on m

The full algebra will be found in Appendix 1, but the principles and main results are presented here.

If all the experimental points were exactly on the line, σ_m would be 0, since we would know the slope exactly. A non-zero value arises from the fact that one or more points deviate from the line.

Each experimental point (x_i, y_i) makes its own contribution to the uncertainty in m. In line with the principle of least squares we are concerned with the sum of the squares of the separate contributions.

Each contribution has two components:

(a) The slope of the best line will change if the y co-ordinate of the typical point changes, as illustrated in Figure 5.6b. The relevant factor is dm/dy_i for the ith point.

(b) The slope will also have to change if the standard deviation σ_{y_i} changes because a larger value produces a smaller weighting factor, which moves the line away from that point, as in Figure 5.6c. Thus the second factor is σ_{y_i}.

Combining these two effects gives the results derived in Appendix 1:

$$\sigma_m^2 = \left[\sum (y - mx - c)^2\right] / \left[N\sum x^2 - (\sum x)^2\right] \tag{32}$$

$$\sigma_c^2 = \left[\sum (y - mx - c)^2\right]\left[\sum x^2\right] / N\left[N\sum x^2 - (\sum x)^2\right] \tag{33}$$

We have all the information necessary to perform these calculations, but it is important to remember that approximations have been made in deriving Equations 32 and 33 (see Appendix 1 for more details).

Cases 2–4

In case 1 the sum to be minimized, shown in Equation 26, treats all the experimental points equally. This is the simplest assumption to make, although we have no evidence one way or the other, since only single measurements are taken at each point. In the other three cases, however, repeated readings yield values of σ_y and in case 4 also σ_x. The values of these uncertainties determine the weight, W_i, which we ascribe to the typical *i*th point. The greater the weight of a particular point, the closer will the line approach it, and vice versa. Although the details differ between the three cases we can make a simple modification to Equation 26 to express the sum to be minimized by the method of least squares. It is

$$S(2-4) = \sum W_i(y_i - mx_i - c)^2 \tag{34}$$

Now we have to derive expressions for W_i for each of the three cases. But before doing that it is worth noting that Equation 34 can also be used to analyze case 1 since the weights would all be (assumed) equal. It is only the relative weights that are significant, rather than their absolute magnitudes, so we could make them all 1, in which case we arrive back at Equation 26.

In case 2 we have $W = (1/\sigma_y)^2$ and we take the simplest case of equal values of σ. This is not the same as case 1, however, since we now have *evidence* for the equal values of σ, while we only assumed it there.

Case 3 allows variation in the values of the uncertainties on the points, though it is still assumed to arise only from the measurements of the y variable. Thus $W_i = (1/\sigma_{y_i})^2$.

Case 4 is more complicated because we have to include error measurements on both the x and y variables. We will be content to quote the formula as

$$W_i = \left[1/\left(\sigma_{y_i}{}^2 + m^2\sigma_{x_i}{}^2\right)\right] \tag{35}$$

A little thought will show that we have a major logical problem here, in that we need to have a value for the slope, m, to calculate W, but we need the latter to calculate the former – an example of the chicken and egg problem. But physicists are nothing if not resourceful, and we solve the problem by using the *method of successive approximations*. That is, we assume an approximate value for m, calculate the W values, use those to calculate a better value for m, and so on until successive values are the same within the required precision. While it is not obvious that the values of m will converge to a fixed value, as opposed to diverging or oscillating, it has been in all the cases I have tried. Clearly the better your starting value for m the quicker you will reach the final value, but there is no significant problem in calculations using modern computers. A summary of my program is appended. Because of the interactive nature of these calculations we cannot write simple expressions for the results of the least squares fit as we can for cases 1–3. These are summarized below:

$$m = \left(\sum W \sum Wxy - \sum Wx \sum Wy\right) / \Delta \tag{36}$$

$$c = \left(\sum Wy \sum Wx^2 - \sum Wx \sum Wxy\right) / \Delta \tag{37}$$

$$\sigma_m = \left(\sum W / \Delta\right)^{1/2} \tag{38}$$

$$\sigma_c = \left(\sum Wx^2 / \Delta\right)^{1/2} \tag{39}$$

where

$$\Delta = \sum W \sum Wx^2 - \left(\sum Wx\right)^2 \tag{40}$$

Combination of errors

Introduction

We have seen how to specify the quality of the measurement of a single variable, by calculating the standard error of the mean. In the absence of systematic errors, the smaller this value the better the observations. We can achieve a small value by taking many repeated measurements, or by careful measurements leading to a smaller standard deviation. But, if we have more than one variable in our experiment how do we combine the errors on each to give the error in the final quantity?

The answer to this question depends upon whether the separate variables are correlated or not. For simplicity let us consider two variables x and y, and assume them to be *not* correlated. The extension to more than two variables, and two which are correlated, will be considered later; let us start with the easiest case first!

Zero correlation means that if x increases then we have no idea whether y increases or decreases. That is, if x changes to $x + dx$ and y changes to $y + dy$, then dy is just as likely to be positive as negative. If we make a number of measurements of x and y then the product $dx \cdot dy$ is just as likely to be positive as negative, and the average will tend to zero.

We have two experimental variables x, y each subject to error, and want to calculate the error on some function of x and y that we will call z, i.e.

$$z = f(x,y) \tag{41}$$

When x changes to $x + \Delta x$ and y changes to $y + \Delta y$ we assume z to change to a value $z + \Delta z$, where

$$z + \Delta z = f(x + \Delta x, y + \Delta y)$$

Subtracting,

$$\Delta z = f(x + \Delta x, y + \Delta y) - f(x,y) \tag{42}$$

Taylor's theorem in two variables can be written in the approximate form

$$f(x + \Delta x, y + \Delta y) = f(x,y) + \Delta x(\partial f/\partial x) + \Delta y(\partial f/\partial y) + \ldots \tag{43}$$

where the higher order terms are assumed negligible. This means that the function is assumed linear in x and y over the range being considered, Δz, which will be reasonable provided Δz is small. The partial differential coefficient $\partial f/\partial x$ is evaluated by differentiating f with respect to x, treating y as if it was a constant.

Thus, approximately,

$$\Delta z = \Delta x(\partial f/\partial x) + \Delta y(\partial f/\partial y) \tag{44}$$

This is not helpful since Δx and Δy are just as likely to be positive as negative, so with a large number of observations Δz tends towards an average of zero. We are looking for a non-zero number to express the spread in the values of z and can achieve this by squaring Δz:

$$(\Delta z)^2 = (\Delta x)^2(\partial f/\partial x)^2 + (\Delta y)^2(\partial f/\partial y)^2 + 2\,\Delta x\,\Delta y(\partial f/\partial x)(\partial f/\partial y) \tag{45}$$

The last term averages to zero provided x and y are not correlated. (If they had a positive correlation then $a + \Delta x$ would correspond to $a + \Delta y$ and $a - \Delta x$ to $a - \Delta y$, and in either case the product would be positive. We will deal with this situation later.)

So the formula we use for the combination of errors is approximately

$$(\Delta z)^2 = (\Delta x)^2(\partial f/\partial x)^2 + (\Delta y)^2(\partial f/\partial y)^2 \qquad (46)$$

provided Δx and Δy are both small and x and y are not correlated.

Applications to various functions

Function	Formula
$x + y$ and $x - y$	$(\Delta z)^2 = (\Delta x)^2 + (\Delta y)^2$
x/y and xy	$(\Delta z/z)^2 = (\Delta x/x)^2 + (\Delta y/y)^2$
x^n	$(\Delta z/z) = n(\Delta x/x)$
$\ln x$ (i.e $\log_e x$)	$\Delta z = \Delta x/x$
$\log_{10} x$	$\Delta z = (1/\ln 10)(\Delta x/x)$
e^x	$\Delta z = e^x \, \Delta x$
$\sin x$	$\Delta z = \cos x(\Delta x)$
$\cos x$	$\Delta z = -\sin x(\Delta x)$

The negative sign in the last formula may need explanation: it simply means that as x increases z decreases. Since we are dealing with random effects here, x is equally likely to be positive as negative, and so is z.

Example 5.6 – error on a difference

Have you ever asked yourself what effect gravity has on the rate at which your heart beats? Some years ago I measured my heart rate, alternately standing and lying flat, and found the following mean values after five repetitions:

$H_S = 58.8 \pm 0.4 \text{ min}^{-1}$
$H_L = 52.6 \pm 0.4 \text{ min}^{-1}$

Taking the difference will show the effect of gravity. The calculation is

$$H_S - H_L = (58.8 - 52.6) \pm [(0.4)^2 + (0.4)^2]^{1/2} \text{ min}^{-1}$$
$$= 6.2 \pm 0.6 \text{ min}^{-1}$$

Since the difference is more than ten times the error, the effect of gravity is very significant.

Another interesting point is brought out by comparing percentage errors on either H_S or H_L and the difference, $H_S - H_L$. These are

Error on $H_S = (0.4/58.8) \times 100 = 0.7\%$
Error on difference $= (0.6/6.2) \times 100 = 10\%$

So the percentage error on the difference is much greater than that on the initial measurements. This will always arise when you are looking for small changes. Since it is easier to look for changes in an experimental variable than its absolute value (see Ch. 3) this situation is quite common in physics. Sometimes this amplification of errors can be avoided by measuring the difference directly, but not in this case as we cannot both stand up and lie down at the same time! If you try the experiment, do not be surprised if your heart rates are higher than mine – I have inherited an efficient heart–lung system and was very fit when the measurements were made.

Example 5.7 – the power law in operation

The viscosity of a liquid is a measure of how readily it flows when subject to a pressure difference. The system may be water driven by a pump round a circuit including a boiler and a number of radiators, or oil acting as a lubricant in an engine. In all cases the viscosity of the liquid must match the needs of the application.

One way of measuring the viscosity, η, of a liquid is to force it under pressure, p, through a narrow tube, of length l and internal radius a, and measure the volume flowing through per second, V. The relevant equation is

$$V = p\pi a^4/8\eta l \tag{47}$$

Unless the liquid is very viscous (in which case you would use a different method anyway!) the radius of the tube must be small. This makes it more difficult to measure than V, p or l, anyhow, but there is the added problem of error multiplication through the combination of errors rule, since

$$\Delta(a^4)/a^4 = 4\,\Delta a/a \tag{48}$$

i.e. the fractional error on a has been multiplied by 4, making it almost certainly the least accurate measurement to be made. Your experimental design must recognize this, and make special provision for measuring the tube radius accurately.

Example 5.8 – diminishing error (sometimes the error gets smaller!)

In many cases, it is true, combining errors magnifies the percentage error; but there are some exceptions. The forward current, I, passing through a semiconductor diode across which a potential difference, V, is maintained is given by

$$I = I_0 e^{bV} \tag{49}$$

where b and I_0 are constants for the particular diode used. This would be transformed, by taking logarithms, so that a straight line graph would be produced:

$$\ln I = \ln I_0 + bV \tag{50}$$

Let us assume that both I and V are measured to 1% accuracy, i.e. a typical pair of values are

$I = 10.0 \pm 0.1$ mA
$V = 1.00 \pm 0.01$ V

For the current we need to know the error on $\ln I$ since this is what is plotted on the graph. It is

$$\Delta(\ln I) = \Delta I/I = 0.1/10 = 0.01$$

and

$$\ln I = \ln(10.0) = 2.30 \tag{51}$$

and so the percentage error on the relevant variable, $\ln I$, has been reduced to $(0.01/2.30) \times 100 = 0.4\%$, compared with 1.0% for I. Instead of congratulating ourselves on this good fortune we ought now to be thinking of increasing the accuracy in the voltage measurement to bring it into line with that for $\ln I$. All this assumes that you become aware of your good fortune before completing the measurements, which leads us to the important subject of experimental design, dealt with in Chapter 3.

Example 5.9 – a computer simulation

I first came across the program "GAUSS", written by Michael Taylor, in the March 1990 edition of BEEBUG, a magazine for users of the BBC microcomputer, and have since added other functions. The program (see Appendix 4) asks you to enter an equation, in the style of BBC BASIC, together with the mean and standard deviation of each of the variables involved (up to a maximum of three). You are then free to choose the number of trials, or "measurements", which you want the computer to make. For each of these the computer chooses a random number for each variable, in the range determined by the standard deviations specified. The formula is then used to calculate a value for the function. As values are accumulated they are used to calculate the mean, standard deviation, and standard error of the mean. Histograms are also plotted for the function and each of the variables used.

The simulation has a number of uses. First, it enables you to check calculations of error combinations in cases that are readily calculated. The measurements of heart rate discussed in Example 1 would be typical. But, whereas in the actual measurements I only made five repetitions, the computer can deliver far better statistics without getting noticeably tired. Hence the 1000 trials shown in Figure 5.7a). A number of comments are appropriate:

(a) In example 1 we used the standard error of the means (0.4 min^{-1}) because we were comparing the *means* of five measurements. Now we are calculating *individual* values of the difference, F, so we need to use the standard deviation. Since there were five measurements, this makes the standard deviation $0.4 \times 5^{1/2} = 0.9$.

(b) The mean value of F, 6.26, is within one and a half standard errors, 0.04, of the expected value of 6.2.

(c) The standard deviation on F, 1.31, is close to the value we would expect for the error on a difference: $(0.92 + 0.92)^{1/2} = 1.27$.

(d) Because we have now made 1000 "measurements" compared with the five real ones, the standard error, 0.041, is much smaller: the expected value is $0.6 \times (5/1000)^{1/2} = 0.042$, another good agreement.

(e) Finally we have a display of the distributions of values for the 1000 trial calculations for X, Y and F. Each looks roughly Gaussian in shape, as we would expect for random fluctuations about the means. There is virtually no overlap between the distributions of X and Y, which confirms that their difference is real. This is further illustrated by the distribution of F values all being above zero.

(f) Figure 5.7b shows a similar calculation, but for 100 000 trials, with the vertical scale adjusted appropriately. Check that the comments made above also apply here.

A second use of the simulation, after you have become confident in the results using simple formulae, allows you to look at more complicated ones, where the algebra is more difficult. An example arises when you use a prism spectrometer

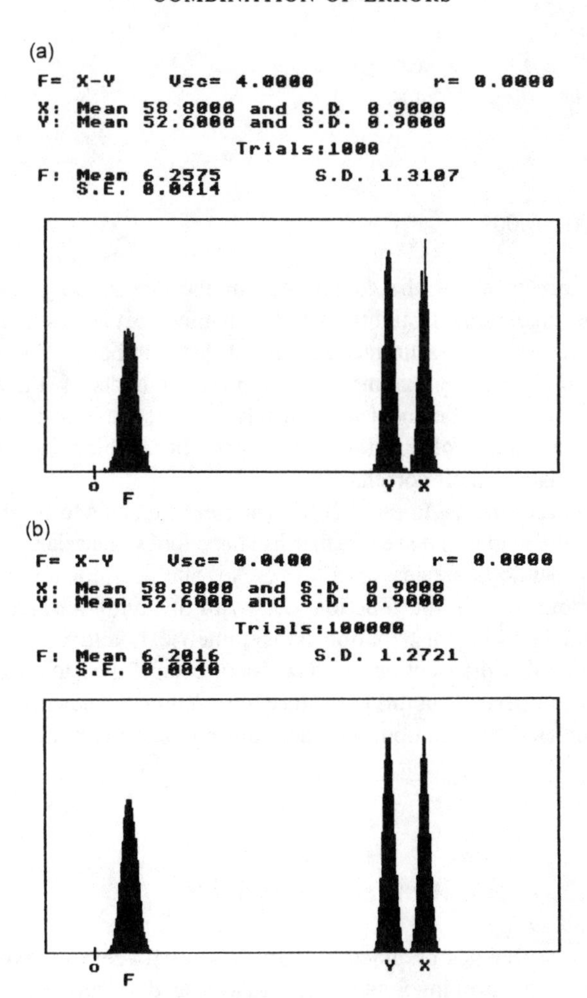

Figure 5.7 Histograms generated by the program GAUSS for (a) 1000 trials and (b) 100 000 trials for the measurements of heart rate discussed in Example 1.

to measure the refractive index of the glass used to make the prism. The formula is

$$\mu = \sin[(A + D)/2]/\sin(A/2) \tag{52}$$

Where A and D are angles measured in the experiment (A is the angle of the prism, and D the angle of minimum deviation of the beam when refracted by the prism). Remembering that the angles must be specified in radians, a typical set of measurements might give the following results:

$$A = 1.047 \pm 0.001 \text{ radians}$$
$$D = 0.649 \pm 0.001 \text{ radians}$$

Thus

$$\mu = 1.500 \pm 0.001$$

You may care to check the consistency of these results by analyzing the equation. The algebra is straightforward, but unwieldy. I have assumed no correlation between the measurements A and D, for simplicity. You would have to check this in practice by making alternate measurements of A and D to estimate the correlation coefficient. Unfortunately, in practice it is simpler to make a series of measurements of A followed by a series of D which does not give any reasonable estimate of their correlation.

Finally, it is rare indeed to make sufficient measurements to be able to plot a distribution of the values and see whether its shape looks Gaussian, for example. The computer can do this every time, of course, and occasionally we can learn important lessons. Using the function $1/X^4$,with X having a value 2.0 ± 0.2, shows that the resulting distribution is asymmetrical, and warns us against always assuming that distributions are Gaussian, even if the input data are. Not that this should surprise us in this case since we would need negative values of F to get a symmetrical distribution, and these are not possible with the conditions stated.

Correlated variables

The discussion so far has assumed no correlation between the two measured variables, so that the term involving $\Delta x\, \Delta y$ has averaged to zero over a number of measurements. But when the variables are correlated this term must be included, yielding an expression for the combined error as follows:

$$(\Delta z)^2 = (\Delta x)^2(\partial f/\partial x)^2 + (\Delta y)^2(\partial f/\partial y)^2 + 2r\,\Delta x\,\Delta y(\partial f/\partial x)(\partial f/\partial y) \quad (53)$$

The first two terms are those used earlier for uncorrelated variables. The final term allows for correlation between the variables and includes the correlation coefficient, r, as a measure of its magnitude. Clearly if $r = 0$ we revert back to the earlier case, while if $r =\ 1$ the second term has its highest effect.

Example 5.10 – correlated variables

As so often in science we look at a new idea with an example that is simple and for which the answer is almost intuitively obvious. In this way we can convince ourselves of the correctness of what we are doing before tackling more complicated problems. We will look at the error on the function x^2, first by seeing it as a single variable raised to the power 2, and then as two separate variables, x and x, multiplied together. Clearly we must get the same answer for the combined error whichever way we approach the function. If we do not then something is wrong.

First approach

$$z = x^2 \tag{54}$$

Since there is only a single variable, x, the idea of correlation is irrelevant. We simply apply the power law formula derived earlier to give:

$$\Delta z / z = 2\Delta x / x$$
$$\therefore \Delta z = 2x\Delta x \tag{55}$$

Second approach

$$z = x \cdot x \tag{56}$$

Here we will be silly and assume there is no correlation between the two variables, x and x, when multiplied together. Although common sense tells us that there is a correlation coefficient of 1 in this case, this will not be so obvious in later situations. We should not be surprised to get the wrong result by using a formula assuming zero correlation:

$$(\Delta z/z)^2 = (\Delta x/x)^2 + (\Delta x/x)^2 \tag{57}$$

or

$$\Delta z = 2^{1/2} x\Delta x \tag{58}$$

Third approach

Here we use the product formula, but include the third term involving the correlation coefficient:

$$(\Delta z)^2 = (\Delta x)^2 x^2 + (\Delta x)^2 x^2 + 2\,\Delta x\,\Delta x\,x \cdot x \tag{59}$$

which gives

$$\Delta z = 2x\,\Delta x \tag{60}$$

This is the result we obtained with the (correct) first approach, but differs from the result obtained by making the false assumption of zero correlation. The lesson is simple: take correlations into account when combining errors, though it is not always easy to find a good value for r.

Example 5.11 – a computer simulation

The original version of the program GAUSS, discussed earlier, did not include the possibility of correlations between the variables. The program now includes my addition to allow for this. In this example I am using the simple function $X \cdot Y$ with values $X = 2.0 \pm 0.2$ and $Y = 4.0 \pm 0.4$ and the extreme range of correlation coefficients −1, 0 and +1. The results are shown in Figure 5.8. They need a little explanation:

$r = -1$ This value of correlation coefficient produces a graph of X against Y that is a straight line with negative slope, so that large values of X are associated with small values of Y, and vice versa. On multiplying, therefore, no high values arise since large values of each variable do not occur together. The top half of the distribution of $X \cdot Y$ is therefore missing and the standard deviation is consequently low.

$r = 0$ The graph is now a scatter diagram, and a full range of values of $X \cdot Y$ is possible. The distributions look a reasonable shape, and the standard deviation is close to what we would expect from the formula derived earlier, without correlation, 1.13.

$r = +1$ The line is of positive slope, so high values of X are associated with high values of Y, and vice versa. On multiplication, both large and small values are obtained, and the distribution is stretched compared with that for $r = 0$. The standard deviation is thus larger.

Use of random numbers

Introduction

When learning to use statistical ideas for the analysis of experimental data it is useful to start with situations for which the result is known. Then you can concentrate on the statistics without worrying whether there is anything wrong with the data. You might think it would be easy to select an experiment in which the measurements were sufficiently under your control to serve the purpose, but systematic errors are more invasive than you might think. Automatic measure-

(a) F= X*Y Vsc= 0.1000 r= -1.0000

X: Mean 2.0000 and S.D. 0.2000
Y: Mean 4.0000 and S.D. 0.4000

Trials:10000

F: Mean 8.0000 S.D. 0.1093
S.E. 0.0011

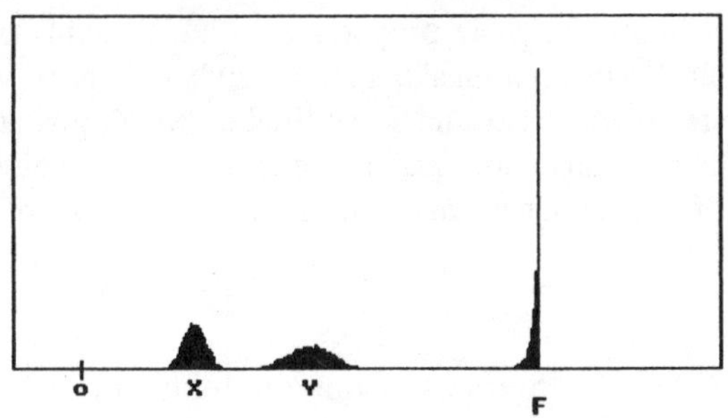

(b) F= X*Y Vsc= 0.6000 r= 0.0000

X: Mean 2.0000 and S.D. 0.2000
Y: Mean 4.0000 and S.D. 0.4000

Trials:10000

F: Mean 7.9808 S.D. 1.1207
S.E. 0.0112

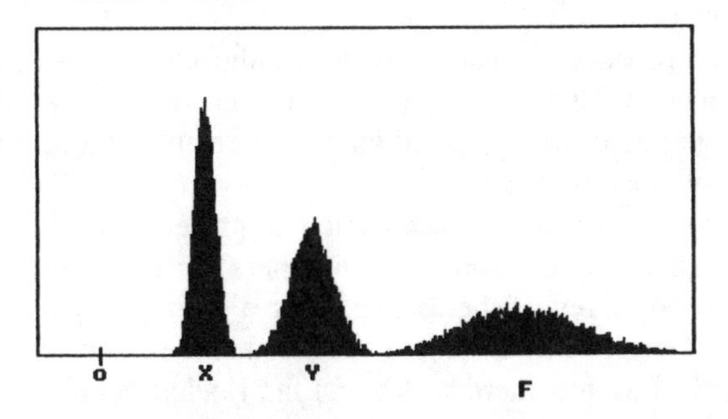

(c) F= X*Y Vsc= 0.6000 r= 1.0000

X: Mean 2.0000 and S.D. 0.2000
Y: Mean 4.0000 and S.D. 0.4000

Trials:10000

F: Mean 7.9760 S.D. 1.5966
S.E. 0.0160

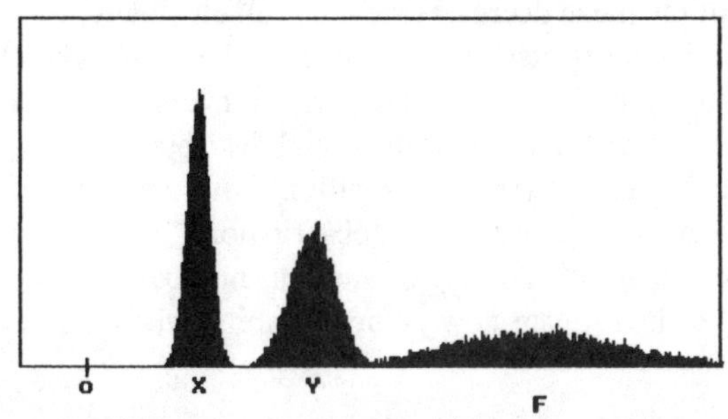

Figure 5.8 Histograms generated by the program GAUSS for the function $X \cdot Y$ (with $X = 2.0 \pm 0.2$ and $Y = 4.0 \pm 0.4$) for the correlation coefficients (a) –1, (b) 0 and (c) +1.

ment achieved by connecting apparatus to a computer will remove human fallibility, but the apparatus will still be subject to variations in temperature, pressure, etc. An alternative is to use random numbers to illustrate the ideas.

The other highly desirable property of random numbers is that there are plenty of them. If you try to make a large number of measurements by reading apparatus yourself you are bound to get tired or bored, giving rise to systematic errors of unknown magnitude. Automatic measurement will again help, but the longer a set of measurements takes, the greater the chance of systematic effects intruding.

Sources of random numbers

(a) Tables of random numbers are available in various books of statistics, for example *Statistics* in *Schaum's outline series*, written by Murray R. Spiegel. This has 1000 random numbers in the range 0–9.
(b) For random numbers in the range 1–6 you can throw a die, assuming it to be unbiased.
(c) Certain processes in nature, such as radioactive decay, are random in that we cannot predict when a particular nucleus will decay. All we know is the average rate of decay, based on measurements of a source consisting of a large number of nuclei.
(d) Any computer worthy of the name will present you with a random number generator. These sequences of numbers are referred to as quasi-random because they repeat themselves after a certain sequence. The repetition interval depends on the length of a word in the computer, and in the case of a word of 32 bits there will be all the random numbers you are likely to need before the cycle repeats itself.

We have already used random numbers in four ways to illustrate statistical ideas:

(a) Figure 5.1 shows how the mean and standard deviation tend to constant values as the number of observations increases. In contrast, the standard error of the mean decreases steadily, albeit slowly.
(b) Figure 5.2 illustrates the central limit theorem, whereby any distribution tends to Gaussian as the number of events added together increases, irrespective of the shape of the initial distribution.
(c) Figure 5.5 illustrates the transition from zero to full correlation as the amount of randomness in the "observations" decreases.
(d) Figure 5.6 shows the use of random numbers in the computer program GAUSS to investigate how errors combine when we measure two or more variables.

Goodness of fit

It was shown earlier in Figure 5.4 how a distribution that starts as rectangular seems to get closer to Gaussian as we add measurements together rather than keeping them separate. It is all very well to suggest trends in this qualitative manner, but scientists must try to be quantitative. We do this by comparing the experimental values, $N_{observed}$, with those predicted by the theory that we are applying to the data, N_{theory}.

You could invent many formulae to bring about this comparison, but the one usually used is that called "chi squared", defined as

$$\chi^2 = \sum\left[(N_{observed} - N_{theory})^2 / N_{theory}\right] \tag{61}$$

where the summation is taken over all the experimental data. It is clear by inspection of this formula that a perfect fit would yield the value $\chi^2 = 0$, but remember that experimental measurements always have random fluctuations, so that a value of χ^2 can be too small to be believable. At the other end of the scale a very poor fit gives a large value of χ^2, though you do not yet know what range of values are to be expected in practice.

The steps to be taken in assessing the goodness of fit between experiment and theory are:

(a) Calculate a value of χ^2 using the equation above.

(b) Make a hypothesis about the fit. A common one is called the *null hypothesis*, which simply means that you assume no difference between theory and experiment.

(c) Look at data such as that contained in Table 5.4, overleaf, showing values of χ^2 for various degrees of freedom, ν, and probabilities, p, that the null hypothesis is true. For example, with $\nu = 5$, you would require a value of χ^2 of 1.15, or less, to be at least 95% confident that the theory and experiment were in agreement.

Example 5.12 – goodness of fit (from Fig. 5.9)

The nature of random numbers leads us to expect the distribution of single values to be rectangular. The top diagram seems to show this is plausible, but this is only qualitative. The value of χ^2 is 1.74. The 6 divisions of the bar chart give $\nu = 5$, so that the probability of the fit being good is a little less than 90%.

The second diagram suggests the fit is poor, and this is confirmed by the value of $\chi^2 = 2130$. Since a value of merely 15.1 is needed for the goodness of fit to be only 1% likely, it is clearly considerably less probable than this. Certainty may not have any part in science but this is as close as you are likely to get.

Table 5.4 χ^2 values.

ν	p 0.01	0.05	0.10	0.50	0.90	0.95	0.99
1	6.63	3.84	2.71	0.455	0.0158	0.0039	0.0002
2	9.21	5.99	4.61	1.39	0.211	0.103	0.0201
3	11.3	7.81	6.25	2.37	0.584	0.352	0.115
4	13.3	9.49	7.78	3.36	1.06	0.711	0.297
5	15.1	11.1	9.24	4.35	1.61	1.15	0.554
6	16.8	12.6	10.6	5.35	2.20	1.64	0.872
7	18.5	14.1	12.0	6.35	2.83	2.17	1.24
8	20.1	15.5	13.4	7.34	3.49	2.73	1.65
9	21.7	16.9	14.7	8.34	4.17	3.33	2.09
10	23.2	18.3	16.0	9.34	4.87	3.94	2.56
11	24.7	19.7	17.3	10.3	5.58	4.57	3.05
12	26.2	21.0	18.5	11.3	6.30	5.23	3.57
13	27.7	22.4	19.8	12.3	7.04	5.89	4.11
14	29.1	23.7	21.1	13.3	7.79	6.57	4.66
15	30.6	25.0	22.3	14.3	8.55	7.26	5.23
16	32.0	26.3	23.5	15.3	9.31	7.96	5.81
17	33.4	27.6	24.8	16.3	10.1	8.67	6.41
18	34.8	28.9	26.0	17.3	10.9	9.39	7.01
19	36.2	30.1	27.2	18.3	11.7	10.1	7.63
20	37.6	31.4	28.4	19.3	12.4	10.9	8.26
21	38.9	32.7	29.6	20.3	13.2	11.6	8.90
22	40.3	33.9	30.8	21.3	14.0	12.3	9.54
23	41.6	35.2	32.0	22.3	14.8	13.1	10.2
24	43.0	36.4	33.2	23.3	15.7	13.8	10.9
25	44.3	37.7	34.4	24.3	16.5	14.6	11.5
26	45.6	38.9	35.6	25.3	17.3	15.4	12.2
27	47.0	40.1	36.7	26.3	18.1	16.2	12.9
28	48.3	41.3	37.9	27.3	18.9	16.9	13.6
29	49.6	42.6	39.1	28.3	19.8	17.7	14.3
30	50.9	43.8	40.3	29.3	20.6	18.5	15.0
40	63.7	55.8	51.8	39.3	29.1	26.5	22.2
50	76.2	67.5	63.2	49.3	37.7	34.8	29.7
60	88.4	79.1	74.4	59.3	46.5	43.2	37.5
70	100.4	90.5	85.5	69.3	55.3	51.7	45.5
80	112.3	101.9	96.6	79.3	64.3	60.4	53.5
90	124.1	113.1	107.6	89.3	73.3	69.1	61.8
100	135.8	124.3	118.5	99.3	82.4	77.9	70.1

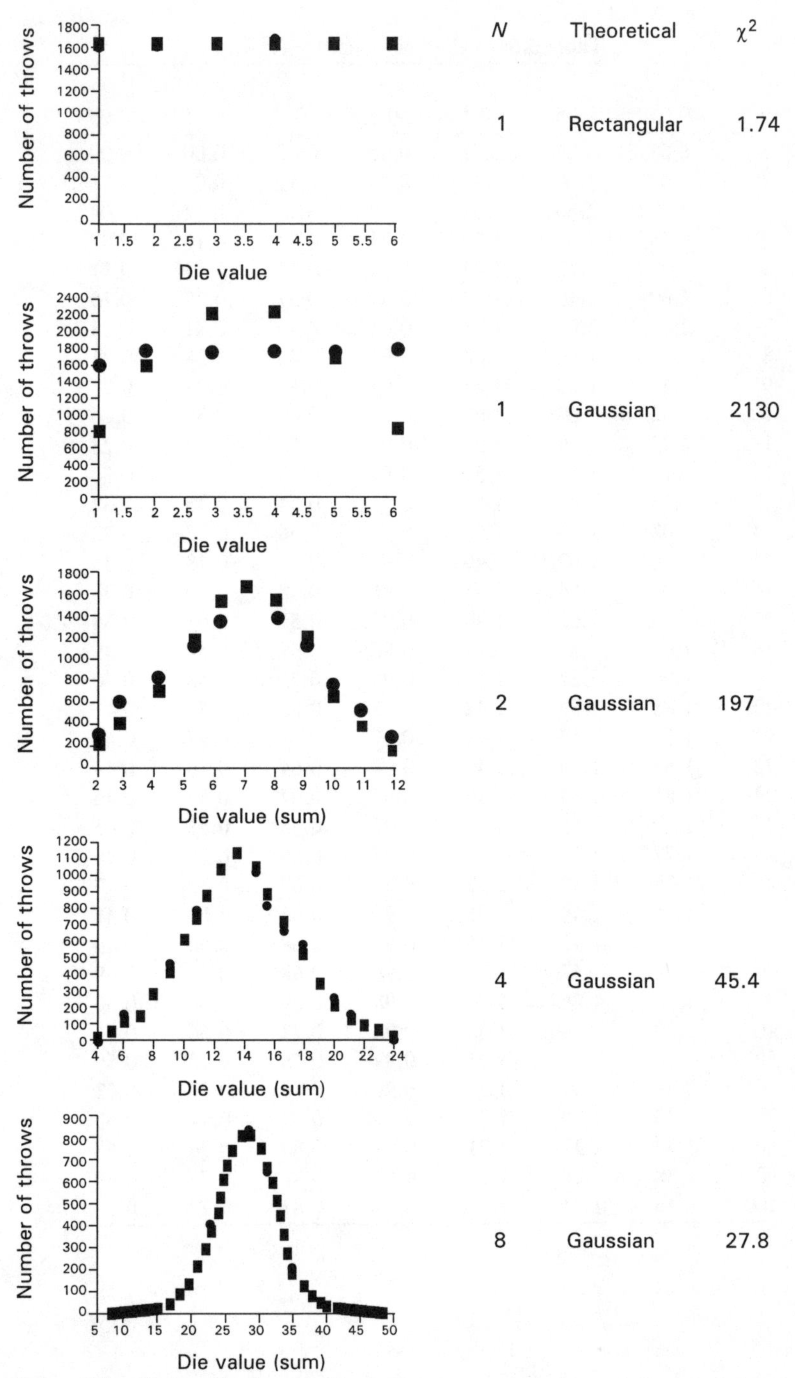

Figure 5.9 Fitting die throws, N at a time, to rectangular and Gaussian distributions. ●, experimental results; ■, theoretical points.

Table 5.5 Values of reduced χ^2 (χ^2/ν).

ν	p 0.01	0.05	0.1	0.5	0.9	0.95	0.99
1	6.63	3.84	2.71	0.46	0.02	0.00	0.00
2	4.61	3.00	2.31	0.70	0.11	0.05	0.01
3	3.77	2.60	2.08	0.79	0.19	0.12	0.04
4	3.33	2.37	1.95	0.84	0.27	0.18	0.07
5	3.02	2.22	1.85	0.87	0.32	0.23	0.11
6	2.80	2.10	1.77	0.89	0.37	0.27	0.15
7	2.64	2.01	1.71	0.91	0.40	0.31	0.18
8	2.51	1.94	1.68	0.92	0.44	0.34	0.21
9	2.41	1.88	1.63	0.93	0.46	0.37	0.23
10	2.32	1.83	1.60	0.93	0.49	0.39	0.26
11	2.25	1.79	1.57	0.94	0.51	0.42	0.28
12	2.18	1.75	1.54	0.94	0.53	0.44	0.30
13	2.13	1.72	1.52	0.95	0.54	0.45	0.32
14	2.08	1.69	1.51	0.95	0.56	0.47	0.33
15	2.04	1.67	1.49	0.95	0.57	0.48	0.35
16	2.00	1.64	1.47	0.96	0.58	0.50	0.36
17	1.96	1.62	1.46	0.96	0.59	0.51	0.38
18	1.93	1.61	1.44	0.96	0.61	0.52	0.39
19	1.91	1.58	1.43	0.96	0.62	0.53	0.40
20	1.88	1.57	1.42	0.97	0.62	0.55	0.41
21	1.85	1.56	1.41	0.97	0.63	0.55	0.42
22	1.83	1.54	1.40	0.97	0.64	0.56	0.43
23	1.81	1.53	1.39	0.97	0.64	0.57	0.44
24	1.79	1.52	1.38	0.97	0.65	0.58	0.45
25	1.77	1.51	1.38	0.97	0.66	0.58	0.46
26	1.75	1.50	1.37	0.97	0.67	0.59	0.47
27	1.74	1.49	1.36	0.97	0.67	0.60	0.48
28	1.73	1.48	1.35	0.98	0.68	0.60	0.49
29	1.71	1.47	1.35	0.98	0.68	0.61	0.49
30	1.70	1.46	1.34	0.98	0.69	0.62	0.50
40	1.59	1.40	1.30	0.98	0.73	0.66	0.56
50	1.52	1.35	1.26	0.99	0.75	0.70	0.59
60	1.47	1.32	1.24	0.99	0.78	0.72	0.63
70	1.43	1.29	1.22	0.99	0.79	0.74	0.65
80	1.40	1.27	1.21	0.99	0.80	0.76	0.67
90	1.38	1.26	1.20	0.99	0.81	0.77	0.69
100	1.36	1.24	1.19	0.99	0.82	0.78	0.70

The last diagram, showing the sums of eight throws of a die, has a value $\chi^2 = 27.8$ for $\nu = 40$, which means that there is a probability of more than 90% that the fit is good. Certainly it is difficult to distinguish the experimental and theoretical points on the graph.

It is not difficult to predict that the fit would get even better, were we to calculate the sums of 16 throws of the die. Try it if you want to practise using a spreadsheet.

Table 5.5 shows values of "reduced chi square", defined as χ^2/ν, which has smaller variation in values than χ^2 itself. In particular, the column with $p = 0.5$ has almost a constant value of about 1. In other words, you can easily remember that a value of the reduced χ^2 of less than 1 is needed for the fit between experiment and theory to be "good".

CHAPTER 6
Models

Introduction

Problems in science are often too complex for us to have a clear grasp of all the details. If, nevertheless, we try to keep on this level of detail it will be difficult for the theory to provide a sound guide on what experiments to do, or how to interpret them when performed. Better to accept the limitations of the human brain in dealing with issues on so large a scale and simplify the problem into the form of a model. The skill lies in making a model that is neither so simple that it is a poor interpretation of the real thing, nor too complex that it is too detailed for practical use. Remember the primary aim of science: to unite theory and experiment.

For many years problems in physics were simplified using linear models, wherein the effect of two variables applied together was simply the sum of the effects of each variable considered separately. This approach, although successful for many problems, never had a chance of explaining phenomena such as weather systems and turbulent motion, which are clearly nonlinear in character. When such problems were tackled with nonlinear models the subject of deterministic chaos was born and was seen to have wide ranging applications in many branches of science.

Sometimes models progress from simplicity to complexity as one's understanding of the problem increases, as in Example 1, below. At other times the model becomes progressively simpler because mistaken ideas introduced in the early versions were later discarded, as in Example 2. The conservative nature of science sometimes allows erroneous ideas to continue longer than one might have expected from our privileged historical standpoint. Such was the case with the use of circular orbits to describe planetary motion, before Kepler's contribution to the evolving model.

Timescales may vary greatly in the development of models. Human beings have speculated, observed, and made models of the operation of the universe for many centuries. The immensity of the problem and the inability to experiment, as opposed to observe phenomena, make this slowness understandable. I shall resist the temptation to contrast this with a model developed over a small timescale, as it is impossible to be sure that any model is complete, and there are plenty of examples when a model was thought to be satisfactory only to require modification later.

The ready availability of computers in recent decades has presented another tool of use in modelling scientific problems. I see no point in simulating situations on a computer if they can be performed perfectly satisfactorily in other ways. If a differential equation can be solved analytically then do so and do not use a computer to produce an approximate numerical solution. Also, if an experiment can be performed with real apparatus then do not waste time making a computer simulation. Rather, use a computer for things it is good at: many repetition of calculations on a problem that has either too many variables or conditions that are very difficult to achieve experimentally. Meteorology is an example of the first, and problems in chaos an example of the second.

Example 6.1 – a model of chaos

The thesaurus on my computer quotes the following synonyms, among others, for the word "chaos": anarchy, disorder, lawlessness, shambles. This is an example where the English language does not encompass the scientific meaning, since "chaos" in scientific usage is none of these things, and is better described as deterministic chaos. It is deterministic in the sense that we could repeat any earlier transformation, were it not for the system being ultra-sensitive to the initial conditions, which we could therefore not reproduce to sufficient accuracy. If, for example, your experiment in chaos was to consist of a forced pendulum, and you wished to repeat a measurement, then you might be able to reset the starting position to a reproducibility of a tenth of a millimetre in a displacement of a few centimetres. Contrast that with the reproducibility achievable with a variable defined by a computer: 1 part in 2^{32} (1 in 4×10^9) say. This is justification enough to introduce the idea of chaos through a computer model, even if it were not invaluable for making the large number of calculations necessary for a full discussion of a problem.

The second important characteristic of a chaotic problem is that it is nonlinear in its response to the experimental variables. This is clear from the example to be considered, which investigates the transformation

$$x_{n+1} = rx_n(1 - x_n) \tag{62}$$

where the term in x_n^2 on the right hand side is the nonlinear contribution. One chooses a value for r (within the range 1–4, though this is not obvious yet), decides on a starting value, x_1, for x_n and calculates the right hand side to give

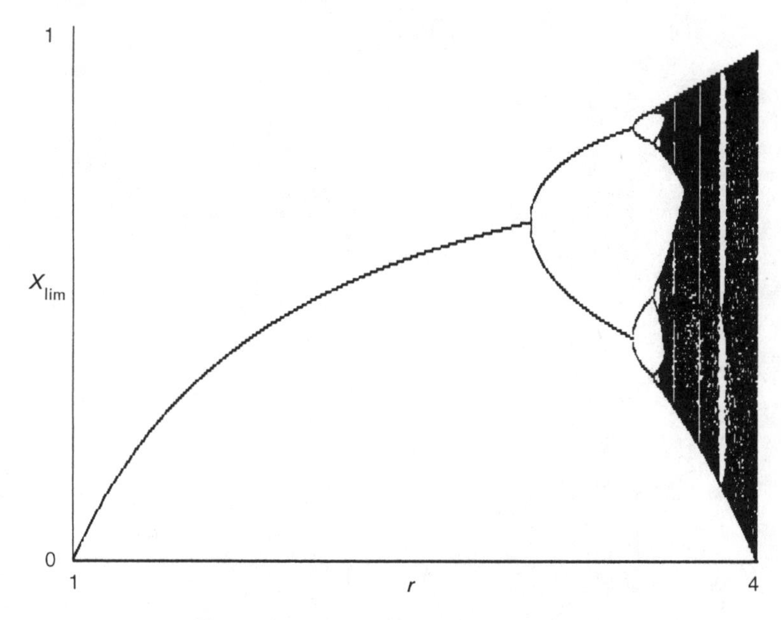

Figure 6.1 Graph of x_{lim} plotted against r.

the next value, x_2, which can then be used on the right hand side again to get the next value, x_3, and so on. This process is called iteration, and is continued until a pattern emerges in consecutive values of x, which I will call x_{lim}, the limiting value(s) of x_n.

This pattern shows up clearly when a graph of x_{lim} is plotted against r (Figure 6.1).

For values of r between 1 and 3 a single value is found for x_{lim} and the graph looks unremarkable in shape. Above $r = 3$ the solution changes in character as it alternates between two values at successive iterations, a process called bifurcation. At higher values of r, further bifurcations occur, with 4, 8, 16, etc., solutions being achieved. At higher values still, the solution becomes chaotic, with many solutions that appear to bear no relationship to each other. However the solutions are reproduced if the calculations are repeated with the same starting value of r.

Figures 6.2 and 6.3 overleaf show an important difference in response to low and high values of r.

In Figure 6.2 the value of $r = 3.2$ is in the region of bifurcation so that a two state solution is obtained. Note that starting values as different as 0.1 and 0.9 make little difference to the final solution, which is reached after only a few iterations.

Contrast this with the situation shown in Figure 6.3 with a value of $r = 3.9$, in the chaotic region. Only the slightest difference in starting value (1 in 10^8) causes the solutions to diverge after less than 100 iterations, showing how

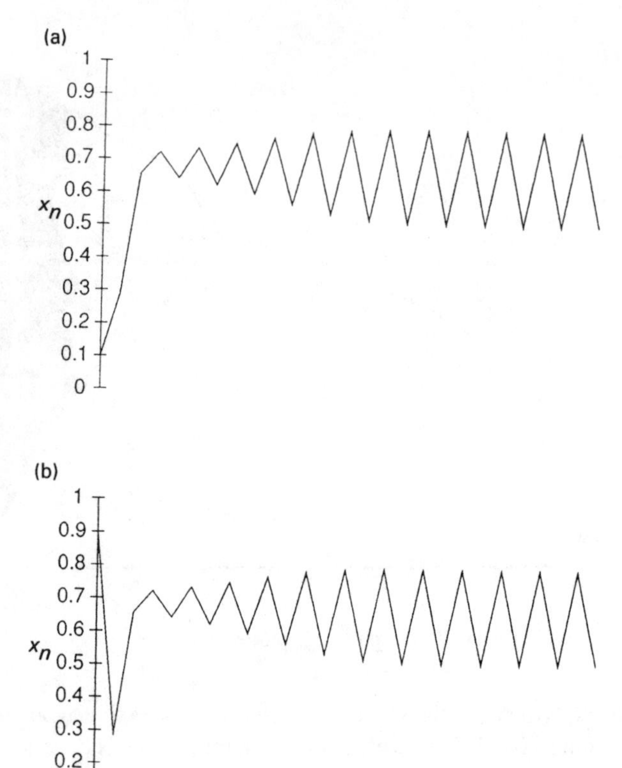

Figure 6.2 Iteration using $r = 3.2$ for (a) $x_1 = 0.1$ and (b) $x_1 = 0.9$.

extremely sensitive chaotic systems are to starting values. If you look carefully at Figures 6.3a,b you will see the solutions to be virtually identical for the first two-thirds of the iterations, but to diverge afterwards. This shows up more clearly in Figure 6.3c, where the difference between the two solutions is plotted.

Example 6.2 – models of motion

This is an example of a model that gets simpler with time, albeit over a very long period. It is an essential part of the scientific process that models of scientific phenomena be tested against experiment if possible, or observation of natural phenomena if not. Failure to do this will probably lead you to unjustified deductions. This was the situation at the time of Aristotle, who thought that a force was needed to keep objects in motion. As you know, if a ball is rolled along a flat

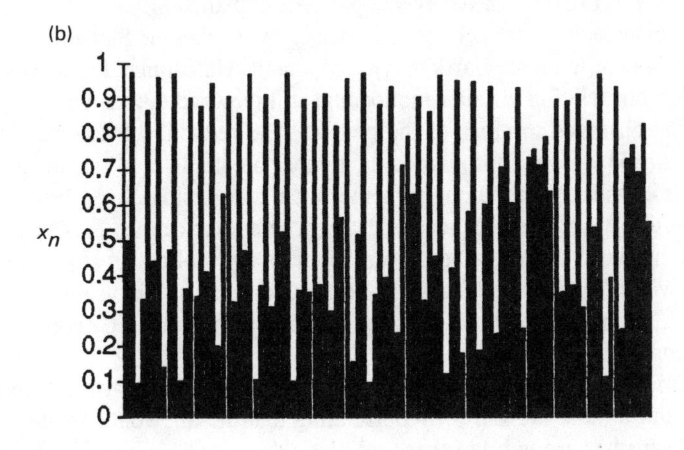

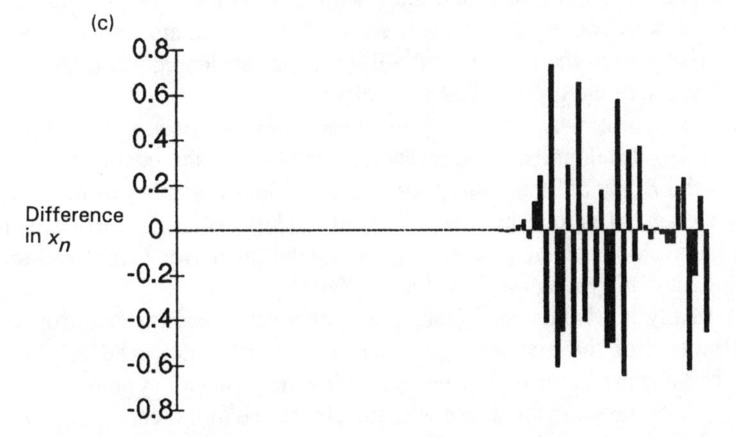

Figure 6.3 Iteration using $r = 3.9$ for (a) $x_1 = 0.5$ and (b) $x_1 = 0.500\,000\,005$. (c) Difference between the solutions in (a) and (b).

surface it will gradually come to rest, and one explanation would be that there is no force to keep it moving. The idea that, in the absence of a force, it would keep moving forever at constant speed in a straight line is by no means obvious. If you are ever tempted to scoff at inadequate models developed in earlier times, remember that you are sitting on giants' shoulders and are privy to the great insight of people like Newton.

Early models of the universe applied general principles based on "perfect" shapes: the circle and sphere. The Earth was normally considered to be at the centre, surrounded by spheres of water, air and fire, with spheres further out on which the Moon, Sun, planets and stars moved. A "prime mover" had to be postulated to produce the required driving force – or the system would presumably have run down like a clockwork toy. The complex motion of the astronomical bodies as seen from the Earth required a complex arrangement of cycles and epicycles, as in Figure 6.4, for even a passable explanation.

Copernicus took the model one step further by placing the Sun at the centre of the Solar System, with the Earth moving around it. Unfortunately, he was unable to abandon his belief in circular motion and still required cycles and epicycles to explain the observed motions of the planets. The reduction from Ptolemy's 80 circles to his 34 was an improvement in number but not in kind. He delayed publishing his ideas for years, though whether from fear of ridicule or the opposition of the church is not clear (echoes of Darwin in a later century?). Certainly the idea that the Earth moved, when it was plain to any fool that it stood still while the Sun, Moon, planets and stars moved around it, must have been revolutionary, even though it had been suggested by some early Greeks, such as Aristarchus.

Copernicus countered Ptolemy's argument that a moving Earth would lead to a great wind around the Earth by postulating that the air would rotate with the Earth. He further argued that there was less chance of his system flying apart due to the motion of the Earth than there would be with Ptolemy's sphere of fixed stars that were set on a much larger scale. The expected apparent motion of the stars, arising from the motion of the Earth, was considered too small to be observed because of the great distances involved.

There is no reason why the history of science should pursue a logical path, and the next big development was not in the model but in the quality of observations. Tycho Brahe did this by repeating observations a number of times and taking an average so as to reduce the effects of random errors. In particular, he recorded the positions of the planets in greater detail than ever before and then passed his data to his young assistant Johann Kepler.

Kepler finally left behind the infatuation with circles, and instead proposed elliptical motion of the planets around the Sun. Furthermore, he combined geometry and algebra in his three laws describing the planetary system:

1. Each planet moves in an ellipse with the Sun at one of its foci.
2. The imaginary line joining the Sun and a planet sweeps out equal areas in equal times.

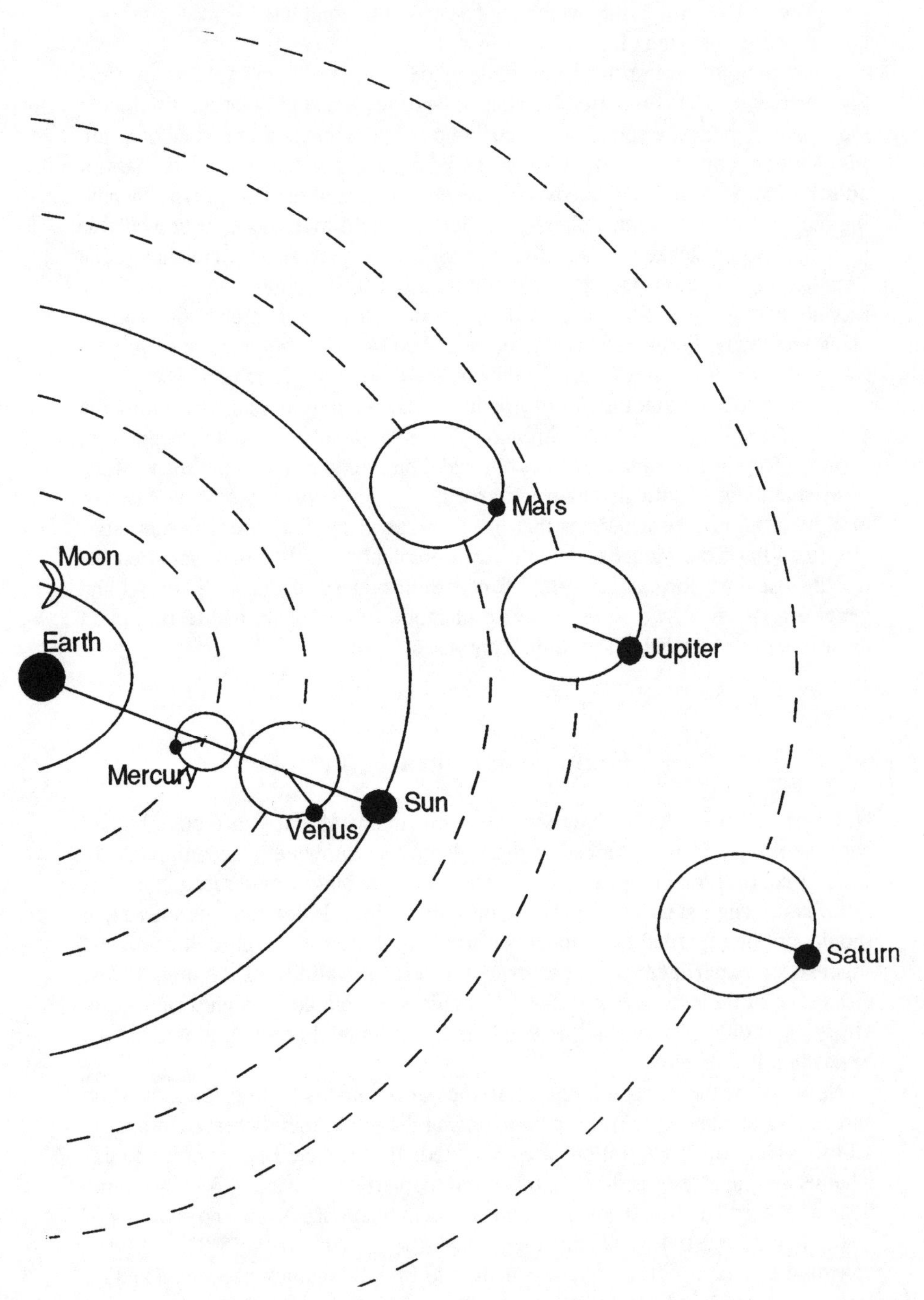

Figure 6.4 Ptolemy's model of the Solar System.

3. The square of the time which a planet takes to complete its orbit is proportional to the cube of its distance from the Sun.

Great though this improvement in the model was, it did not explain *why* these laws were valid. That required Newton's law of universal gravitation. But before then Galileo made the significant contribution of showing that the speed of a falling body is proportional to the time of its fall, and independent of its mass or constitution. He did this by the beautifully simple idea of diluting gravity by having balls roll down inclined planes in order that the timing mechanism available to him, a water clock, was sensitive enough for the purpose. He concluded that force *changes* motion, rather than being required for its continuance. As if that was not enough for a lifetime's work, he made many astronomical observations using the newly discovered telescope, described the parabolic motion of projectiles and made important measurements on the motion of pendulums.

Newton showed that the law of gravity applied equally to planets orbiting the Sun as to apples falling to the ground. The inverse square law followed from applying Kepler's third law to the orbit, and then he deduced that the shape of the orbit should be elliptical, again in agreement with Kepler. His three laws of motion summarized principles that have universal application. This immense clarification of the model of motion, astronomical or earthbound, satisfied the world of science for centuries until Einstein introduced his ideas of special and general relativity. And so the practice of modifying models, and testing them against experiment and observation, continues.

Example 6.3 – heart rate

Not only can a model become simpler with time, but also more complicated because the problem is looked at more deeply as knowledge accumulates. In Chapter 5 I proposed that the effect of gravity on my body would result in a pulse rate faster when standing up than lying down. Simple though this idea is, it should still be regarded as a model because it simplifies a complex situation and suggests an experiment to be performed to test the validity of the model. The difference in the two rates was 6.2 ± 0.6 pulses per minute, a highly significant effect. We could rest the case here and look for the next problem to interest us, or develop it further.

Now one of the delightful characteristics of science is that posing a question and getting an answer to it, instead of closing the book, merely serves to open it a little wider: further questions are prompted. In this case I am tempted to ask whether my heart rate is directly related to the vertical distance (d) between my head and feet. The two readings I have at the moment are prone and standing: $d = 0$, $H = 52.6 \pm 0.4$ beats per minute, and $d = 1.72$ m, $H = 58.8 \pm 0.4$ beats per minute. This is not enough information to find the relation between d and H, since only a straight line can be fitted to two points, and who is to say the relation should be linear?

Another reading could be achieved simply by sitting down, or I could go to greater trouble and lie on a board that could be inclined at various angles to the vertical to get a range of readings. A sensible approach in experiments is to do that which is easy first, and if that result appears promising to make more elaborate measurements later. If an inclined plane and an assistant is readily available then you might decide to go for a full set of readings straight away. It depends on your temperament and what facilities are available.

Since I had not taken readings sitting down at the earlier time, I thought it would be prudent to take a fresh set of all three measurements in pulses per minute:

	Lying	Sitting	Standing
	52	53	58
	51	53	59
	51	52	58
	51	53	59
	52	52	58
Mean	51.4	52.6	58.4
SEM	0.2	0.2	0.2

A scientist tries to be dispassionate about data, but I could not suppress a feeling of surprise and pleasure that the values were close to those measured years earlier, when I was not only younger but fitter. The vertical distance between my head and feet when sitting is 1.24 m, so I can plot a graph of the three values (Fig. 6.5).

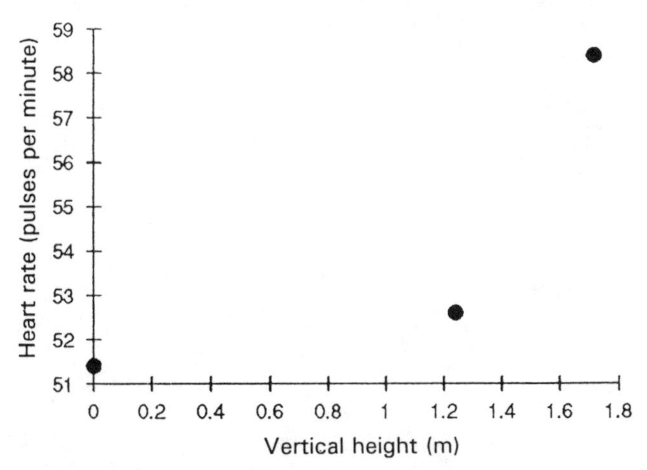

Figure 6.5 Heart rate versus vertical height between my head and feet when lying down, sitting and standing.

Clearly the graph is not linear. Fitting a curve to the graph as a guidance for further measurements could produce a result such as

$$P = 49.5 + 6.36(d - 0.54)^2 \text{ pulses per minute}$$

(Check that this equation gives the measured values of P at the three values of d used.) This would lead us to believe that there should be a minimum pulse rate of 49.5 pulses per minute at a separation between my head and feet of 0.54 m. If I believed this result then I should presumably tilt my bed to the appropriate angle to give the heart maximum rest when asleep. It is much more likely that my model has been stretched beyond the point of usefulness and I had better return from the blind alley into which it has lured me. This is the fate of models when experimental data fail to support their predictions.

A little thought tells us that gravity is not the only variable affecting the heart rate, as extra muscular activity is needed when standing compared with lying or sitting down. To disentangle the two effects we need either to support the body when vertical, or create various stages of muscular activity with the body in one of the three positions. Continue to develop the model for yourself if the subject interests you.

APPENDIX 1

Uncertainty on the line of least squares fit

If all the experimental points were exactly on the line, σ_m would be 0, since we would know the slope exactly. A non-zero value arises from the fact that one or more points deviate from the line. Each experimental point (x_i, y_i) makes its own contribution to the uncertainty in m. In line with the principle of least squares we are concerned with the sum of the squares of the separate contributions.

Each contribution has two components:

(a) The slope of the best line will change if the y co-ordinate of the typical point changes, as illustrated in Figure 5.6b. The relevant factor is dm/dy_i for the ith point.

(b) The slope will also have to change if the standard deviation σy_i changes because a larger value produces a smaller weighting factor, which moves the line away from that point, as in Figure 5.6c. Thus the second factor is σ_{y_i}.

Combining these two effects gives

$$\sigma_m{}^2 = \sum (dm/dy_i)^2 \sigma_{y_i}{}^2 \tag{63}$$

where the sum is to be evaluated over all points on the graph.

But there is a problem! For case 1 we do not know the values of σ_y because we only take a single reading of y at each value of x. In spite of this ignorance we will make the simplifying assumption that σ_y has the same value at each point.

In Equation 30, p. 83 we see that the denominator is not a function of y so we can write it in the simpler form

$$m = \left(N\sum xy - \sum x \sum y\right) / \Delta$$

where $$\Delta = N\sum x^2 - \left(\sum x\right)^2$$

therefore $$dm / dy_i = d\left(N\sum xy / \Delta - \sum x \sum y / \Delta\right) / dy_i \tag{64}$$

For the N terms in each summation, only the ones which include y_i give a non-zero contribution to the differential coefficients. Terms such as y_1 and x_1y_1 derive from measurements that are not related to those from other points on the graph, so give zero contributions to the differentials. We are thus left with the result

$$dm / dy_i = Nx_i / \Delta - \sum x_i / \Delta$$

so $$(dm / dy_i)^2 = \left[N^2x_i^2 - 2Nx_i\sum x_i + (\sum x_i)^2\right] / \Delta^2 \tag{65}$$

substituting this expression into Equation 63 and summing over all N values gives

$$\sigma_m^2 = \sigma_y^2\left[N^2\sum x^2 - 2N\sum x\sum x + N(\sum x)^2\right] / \Delta^2$$

$$= \sigma_y^2\left[N^2\sum x^2 - N(\sum x)^2\right] / \Delta^2$$

$$= \sigma_y^2[N\Delta] / \Delta^2$$

$$= N\sigma_y^2 / \Delta \tag{66}$$

This is a simple enough expression, but we cannot use it because in case 1 we have no knowledge of σ_y. We assumed for simplicity that all the values were equal, but of what magnitude?

To progress, you have to understand the philosophy of physicists. It is simply that we would rather have an approximate answer to a problem than no answer at all. After all, science is concerned with probabilities rather than certainties, so we are never going to have anything but approximate answers; the problems we study are usually too difficult. This does not excuse us from recognizing and clearly stating the approximations we make.

In this case we assume that a point being further from the best line than another point is due to the greater uncertainty in its y measurement, σ_y. We can thus make the following approximation for the distance between point i and the best line:

$$\sigma_y \approx y_i - mx_i - c \tag{67}$$

Averaging over all N points:

$$\sigma_y^2 = \left[\sum (y_i - mx_i - c)^2\right] / N \tag{68}$$

Substituting into Equation 66 and cancelling the Ns gives

$$\sigma_m^2 = \left[\sum (y - mx - c)^2\right] / \left[N\sum x^2 - \left(\sum x\right)^2\right] \tag{69}$$

We have all the information needed to make this calculation, but do not forget the approximations made in its derivation.

A similar procedure is adopted to calculate σ_c:

$$\sigma_c^2 = \sum (dc / dy_i)^2 \sigma_{y_i}^2 \tag{70}$$

where

$$dc / dy_i = d\left(\sum y \sum x^2 / \Delta - \sum x \sum xy / \Delta\right) / dy_i$$

As before, we recognize that the only terms of consequence in the summations are those involving y_i since the rest differentiate to zero. Therefore

$$(dc / dy_i)^2 = \left[\left(\sum x^2\right)^2 - 2x_i \sum x \sum x^2 + x_i^2 \left(\sum x\right)^2\right] / \Delta^2 \tag{71}$$

Substituting into Equation 70 and summing over all N values gives

$$\begin{aligned}
\sigma_c^2 &= \sigma_y^2 \left[N\left(\sum x^2\right)^2 - 2\sum x \sum x \sum x^2 + \sum x^2 \left(\sum x\right)^2\right] / \Delta^2 \\
&= \sigma_y^2 \left[N\left(\sum x^2\right)^2 - \left(\sum x\right)^2 \sum x^2\right] / \Delta^2 \\
&= \sigma_y^2 \sum x^2 \left[N\sum x^2 - \left(\sum x\right)^2\right] / \Delta^2 \\
&= \sigma_y^2 \sum x^2 / \Delta^2 \\
&= \sum x^2 \left[\sum (y - mx - c)^2\right] / N\Delta
\end{aligned} \tag{72}$$

Again we have all the information needed to make a calculation, but remember its approximate nature.

APPENDIX 2

Coefficient of linear correlation r

This is a measure of how close data, representing two variables x and y, are to a straight line. Extreme values are:

(a) 0 for no correlation, i.e. either x or y formed from a set of random numbers;

(b) +1 for perfect positive correlation, i.e. the variables x and y are on a perfect straight line with positive slope;

(c) −1 for perfect negative correlation and negative slope.

The equation defining the coefficient is

$$r = \frac{\sum (x - x_{\text{mean}})(y - y_{\text{mean}})}{\left[\sum (x - x_{\text{mean}})^2 \sum (y - y_{\text{mean}})^2\right]^{1/2}} \tag{73}$$

where x_{mean} and y_{mean} are the mean values.

If the values of y are generated by a random number generator, as in Figure 5.5a, the numerator tends to zero since terms in the summation are equally likely to be positive or negative. The coefficient is slightly different from zero (0.0035 in this case) because of the finite number of points in the data (17).

Using a simple data set such as (1, 1) (2, 2) and (3, 3) it is easy to show that the coefficient is +1. Similarly, a data set consisting of (1, 3) (2, 2) and (3, 1) produces a coefficient of −1.

The calculation of r is included in my program WLSF for fitting the best straight line to data (see Appendix 4).

APPENDIX 3

Books you might enjoy reading

Clark, R. W. 1982. *Freud - the man and the cause.* London: Paladin Books.

Close, F. 1992. *Too hot to handle.* London: Penguin.

Desmond, A. & J. Moore 1991. *Darwin.* London: Michael Joseph.

Goodchild, P. 1980. *J Robert Oppenheimer - shatterer of worlds.* London: BBC.

Harre, R. 1981. *Great scientific experiments.* Oxford: Phaidon.

Feynman, R. P. 1986. *Surely you're joking, Mr Feynman!* London: Unwin Hyman.

Feynman, R. P. 1990. *What do you care what other people think?* London: Unwin Hyman.

Gratzer, W. (ed.) 1989. *The Longman literary companion to science.* Harlow, England: Longman.

Hodges, A. 1992. *Alan Turing - the enigma.* London: Vintage.

Lipson, H. S. 1968. *The great experiments in physics.* Harlow, England: Oliver and Boyd.

Medawar, P. Anything he has written!

Reid, R. 1978. *Marie Curie.* London: Paladin.

Shamos, M. H. 1965. *Great experiments in physics.* London: Holt Rinehart and Winston.

Siegel, A. F. 1988. *Statistics and data analysis.* Chichester, England: John Wiley.

Spiegel, M. R. 1972. *Statistics.* Schaum's Outline series. Maidenhead, Berkshire: McGraw-Hill.

Woodham-Smith, C. 1982. *Florence Nightingale.* London: Constable.

APPENDIX 4

Programs discussed in text

The programs were all written in BBC BASIC. Not being a universally accepted language it was thought undesirable to provide line-for-line listings. Instead, the outline of the programs is shown through a set of procedures that could be written in a form suitable for a particular computer.

Program 4.1: DICE

The space bar is used to select random numbers in the range 1–6 up to selected maximum of T. Values of running mean, standard deviation and standard error are calculated at each stage. The screen displays both the calculated data and a control chart, to compare a long-term mean of 3.5 with current values. An option is available to dump the output to a printer.

```
PROCexplanation
PROCinput
PROCinitialize
PROCprepare-screen
FOR N = 1 TO T
   PROCcalculate
   PROCdata-to-screen
   IF Space-bar pressed THEN NEXT N ELSE END
PROCprinter-option
END
```

Program 4.2: REACTIM

This measures and analyzes your reaction time to a visual stimulus. You may choose between looking at data consisting of ten good stored points, ten bad stored points, or making your own measurements.

```
IF      A$ = "G" THEN PROCread-good-data
ELSE IF A$ = "B" THEN PROCread-bad-data
ELSE IF A$ = "M" THEN PROCmeasure-reaction-time
PROCrunning-calculations
PROCdisplay-menu
  1. PROCtable
  2. PROCtime-series
  3. PROCprinter-option
  4. END
```

Program 4.3: WLSF

This calculates the best straight line for one of the four cases:

Errors on X, Y not known
Errors on Y all equal
Errors on Y not equal
Errors on both X and Y

Menus are provided for selection of case, calculations performed, and type of display required.

```
PROCcase-menu
PROCinitialize
PROCinput-data
PROCdisplay-menu
  1. List input data
  2. Delete selected data
  3. Change input data
  4. Print data and result
  5. New set of data
  6. Calculate best line
  7. Show graph on screen
  8. Dump graph to printer
  9. End program
```

Program 4.4: GAUSS

Various inputs are required to enable calculations and displays to be produced showing the way in which experimental variables combine, with or without correlation. Random values of each variable are produced within the ranges specified by the input data. These are combined according to the formula specified, and the calculations repeated a fixed number of times or until a suitable display is produced. The running values and developing display are shown on screen, with a facility for dumping to a printer if desired.

```
PROCvariables
PROCinitial-values
PROCpicture
REPEAT
   PROCcalculate
   PROCdisplay
   PROCstatistics
UNTIL calculations complete
PROCprinter-option
END
```

Index